C

W
on
Earth?

C000179737

The world in your pocket

Harper Collins*Publishers*
Westerhill Road, Bishopbriggs, Glasgow, G64 2QT

www.collins.co.uk

First published 2006

ISBN-13 978-0-00-721942-1
ISBN-10 0-00-721942-3

Copyright © HarperCollins*Publishers* 2006
Maps © Collins Bartholomew Ltd 2006

All rights reserved. Collins Gem® is a registered trademark of HarperCollins*Publishers* Ltd

The contents of this edition of the Collins Gem Where on Earth? are believed correct at the time of printing. Nevertheless the publisher can accept no responsibility for errors or omissions, changes in the detail given, or for any expense or loss thereby caused.

Printed in Italy by Amadeus S.r.l.

All mapping in this book is generated from Collins Bartholomew digital databases. Collins Bartholomew, the UK's leading independent geographical information supplier, can provide a digital, custom, and premium mapping service to a variety of markets.
For further information:
tel: +44 (0) 141 306 3752
e-mail: collinsbartholomew@harpercollins.co.uk
Visit our website at: www.collinsbartholomew.com

Foreword

Have you ever tried to find the population of Europe, the longest river in Asia or the size of the Atlantic Ocean? Key facts such as these, which are usually buried in larger books, can be found easily and quickly here in *Gem Where on Earth?*

Gem Where on Earth? covers a vast range of physical and human geographical information, including all types of extremes, from the largest, smallest, longest and shortest, to the hottest, coldest, richest and poorest.

This invaluable book contains maps, tables and graphs providing vital information on the physical and political world, the oceans, the world's countries and territories, and global issues including population and cities, climate, the environment, wealth and social indicators, terrorism and travel. Together with a useful web directory, *Gem Where on Earth?* will be invaluable for everyone from trivia buffs and quiz compilers to students and journalists, for ready reference or study, for quizzes or just for peace of mind.

CONTENTS

THE PHYSICAL WORLD

THE SOLAR SYSTEM

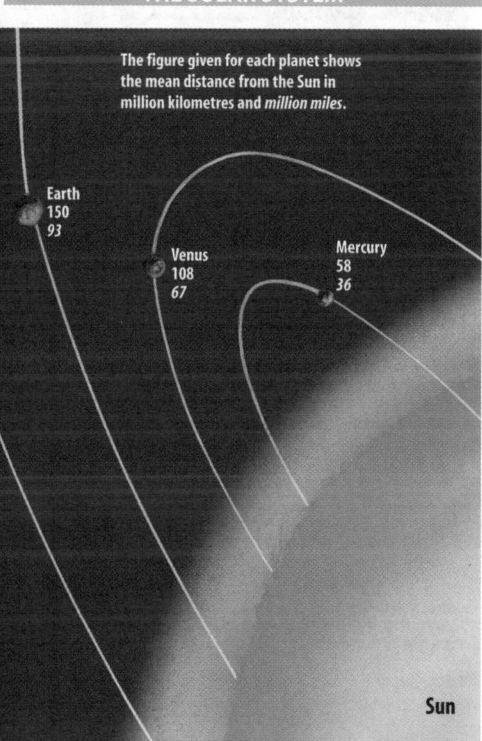

THE PLANETS

	Mercury	Venus	Earth
Mean orbital distance from Sun (million km)	57.9	108.2	149.6
(million miles)	*36.0*	*67.2*	*93.0*
Equatorial diameter (km)	4 879.4	12 103.6	12 756.3
(miles)	*3 032.1*	*7 521.2*	*7 926.8*
Rotation period (earth days)	58.65	-243	23hr 56m 4s
Year (earth days/years)	88 days	224.7 days	365.24 days
Mean orbital velocity (km/s)	47.87	35.02	29.79
(miles/s)	*29.75*	*21.76*	*18.51*
Mean surface temperature (°C)	167	457	15–20
Mass (Earth=1)	0.055	0.815	1 (6×10^{24})
Density (Water=1)	5.43	5.24	5.52
Surface gravity (Earth=1)	0.38	0.91	1
Total number of known moons	0	0	1

Mars	Jupiter	Saturn	Uranus	Neptune	Pluto
227.9	778.4	1 426.7	2 871.0	4 498.3	5 906.4
141.6	*483.7*	*886.6*	*1 784.0*	*2 795.2*	*3 670.2*
6 794	142 984	120 536	51 118	49 528	2 390
4 221.8	*88 850.3*	*7 4901.1*	*31 764.7*	*30 776.7*	*1 485.2*
1.03	0.41	0.44	-0.72	0.67	-6.39
687 days	11.86 years	29.42 years	83.8 years	163.8 years	248 years
24.13	13.07	9.67	6.84	5.48	4.75
14.99	*8.12*	*6.01*	*4.25*	*3.41*	*2.95*
-90– -5	-108	-139	-197	-200	-215.2
0.107	317.82	95.161	14.371	17.147	0.002
3.94	1.33	0.70	1.30	1.76	1.10
0.38	2.53	1.07	0.90	1.14	0.06
2	63	30	21	8	1

THE EARTH

EARTH'S STRUCTURE

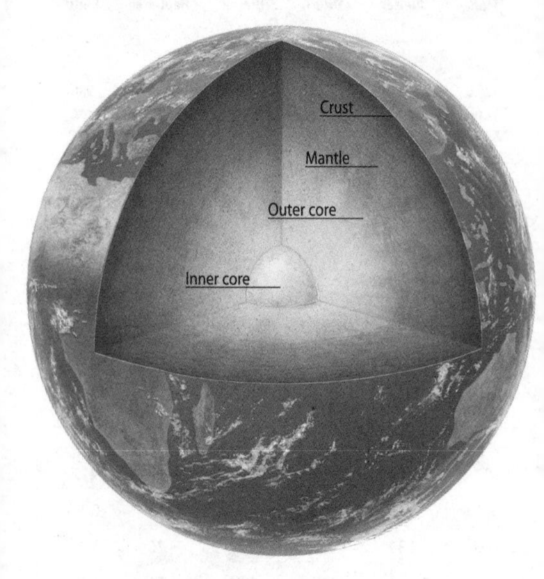

Crust : Average 33 km (20.5 miles) thick
Mantle : 2 850 km (1 771 miles) thick
Outer core : 2 250 km (1 398 miles) thick
Inner core : Radius 1 220 km (758 miles)

EARTH'S DIMENSIONS

EARTH'S DIMENSIONS	
Mass	5.974 X 10^{21} tonnes
Total area	509 450 000 sq km / 196 672 000 sq miles
Land area	149 450 000 sq km / 57 688 000 sq miles
Water area	360 000 000 sq km / 138 984 000 sq miles
Volume	1 083 207 X 10^6 cubic km / 259 875 X 10^6 cubic miles
Equatorial diameter	12 756 km / 7 926 miles
Polar diameter	12 714 km / 7 900 miles
Equatorial circumference	40 075 km / 24 903 miles
Meridional circumference	40 008 km / 24 861 miles

EARTHQUAKES AND VOLCANOES

THE EARTH'S TECTONIC PLATES

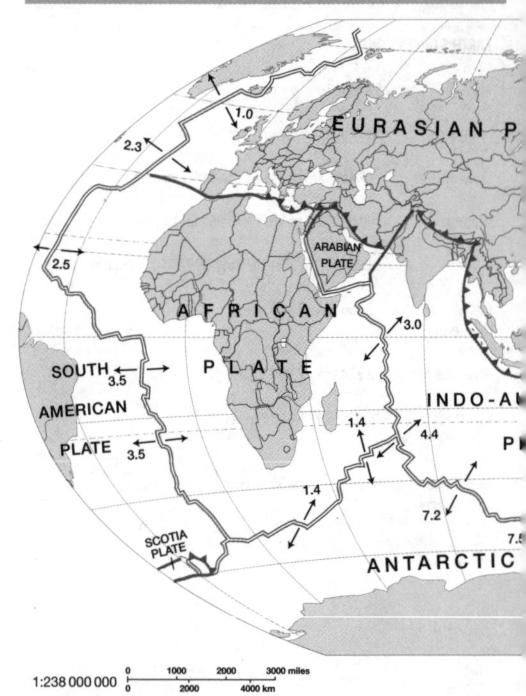

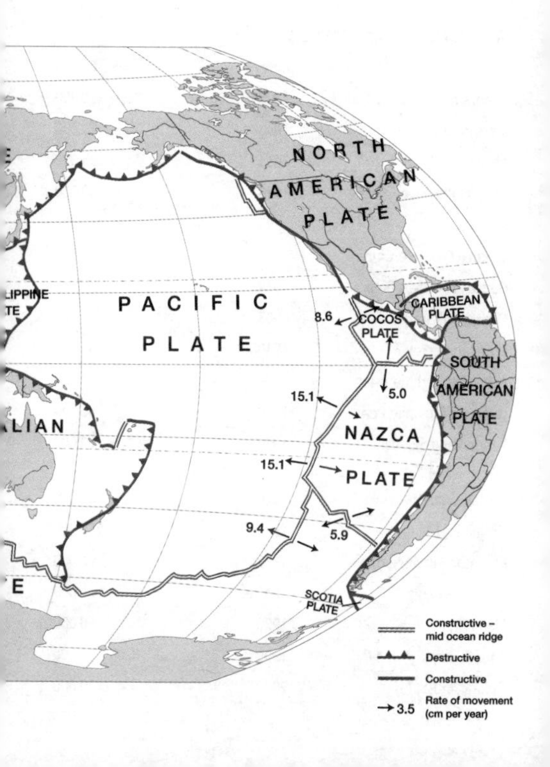
NORTH AMERICAN PLATE
PACIFIC PLATE
CARIBBEAN PLATE
COCOS PLATE
SOUTH AMERICAN PLATE
NAZCA PLATE
SCOTIA PLATE
LIPPINE
TE
LIAN
E
8.6
15.1
5.0
15.1
9.4
5.9
Constructive – mid ocean ridge
Destructive
Constructive
3.5
Rate of movement (cm per year)

DEADLIEST EARTHQUAKES AND VOLCANOES

DEADLIEST EARTHQUAKES 1900–2005

	Date	Deaths
Kangra, India	1905	19 000
Messina, Italy	1908	110 000
Abruzzo, Italy	1915	35 000
Ningxia Province, China	1920	200 000
Tōkyō, Japan	1923	142 807
Qinghai Province, China	1927	200 000
Gansu Province, China	1932	70 000
Quetta, Pakistan	1935	30 000
Erzincan, Turkey	1939	32 700
Ashgabat, Turkmenistan	1948	19 800
Huánuco Province, Peru	1970	66 794
Yunnan and Sichuan Provinces, China	1974	20 000
central Guatemala	1976	22 778
Tangshan, China	1976	255 000
Khorāsan Province, Iran	1978	20 000
Spitak, Armenia	1988	25 000
Manjil, Iran	1990	50 000
Kocaeli (İzmit), Turkey	1999	17 000
Gujarat, India	2001	20 000

Bam, Iran	2003	26 271
off Sumatra, Indian Ocean	2004	>250 000
Northwest Pakistan	2005	87 000
MAJOR VOLCANIC ERUPTIONS 1980–2005		
	Location	**Date**
Mt St Helens	USA	1980
El Chichónal	Mexico	1982
Gunung Galunggung	Indonesia	1982
Kilauea	Hawaii	1983
Ō-yama	Japan	1983
Nevado del Ruiz	Colombia	1985
Mt Pinatubo	Philippines	1991
Unzen-dake	Japan	1991
Mayon	Philippines	1993
Galeras	Colombia	1993
Volcán Llaima	Chile	1994
Rabaul	Papua New Guinea	1994
Soufrière Hills	Montserrat	1997
Hekla	Iceland	2000
Mt Etna	Italy	2001
Nyiragongo	Democratic Republic of the Congo	2002

MOST POWERFUL EARTHQUAKES

Location	Magnitude	Date
Chile	9.5	22 May 1960
Prince William Sound, Alaska	9.2	28 March 1964
off the west coast of northern Sumatra, Indian Ocean	9.0	26 December 2004
Kamchatka, Russian Federation	9.0	4 November 1952
off the coast of Ecuador	8.8	31 January 1906
Northern Sumatra, Indonesia	8.7	28 March 2005
Rat Islands, Alaska	8.7	4 February 1965
Andreanof Islands, Alaska	8.6	9 March 1957
Assam - Tibet	8.6	15 August 1950
Kuril Islands, Russian Federation	8.5	13 October 1963
Banda Sea, Indonesia	8.5	1 February 1938
Kamchatka, Russian Federation	8.5	3 February 1923

THE RICHTER SCALE

The magnitude of earthquakes is measured on the Richter Scale. The scale is logarithmic - an increase in magnitude of 1 unit corresponds to a tenfold increase in the size of an earthquake.

Magnitude	Characteristics
<3.5	Only detected by seismograph
3.5	Only noticed by people at rest
4.2	Similar to vibrations from large vehicle
4.5	Felt indoors; rocks parked vehicles
4.8	Generally felt; awakens sleepers
5.4	Trees sway; causes some damage
6.1	Causes general alarm; building walls crack
6.5	Walls collapse
6.9	Some houses collapse; cracks appear in ground
7.3	Buildings destroyed; rails buckle
8.1	Most buildings destroyed; landslides
>8.1	Total destruction of area

PHYSICAL FEATURES AND EXTREMES

THE CONTINENTS

EUROPE
Area:
9 908 599 sq km
3 825 731 sq miles

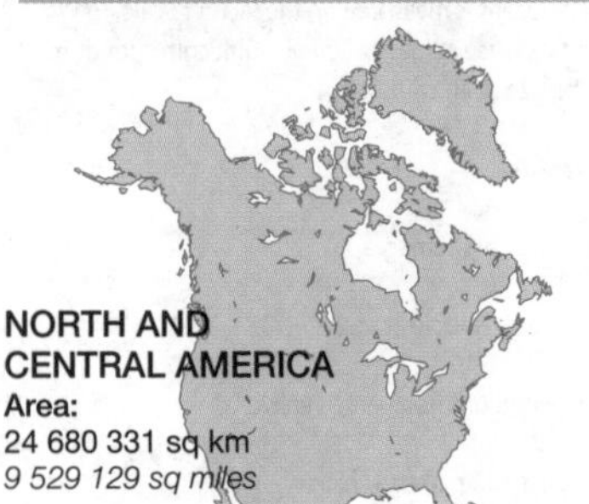

SOUTH AMERICA
Area:
17 815 420 sq km
6 878 572 sq miles

ANTARCTICA
Area:
12 093 000 sq km
4 669 133 sq miles

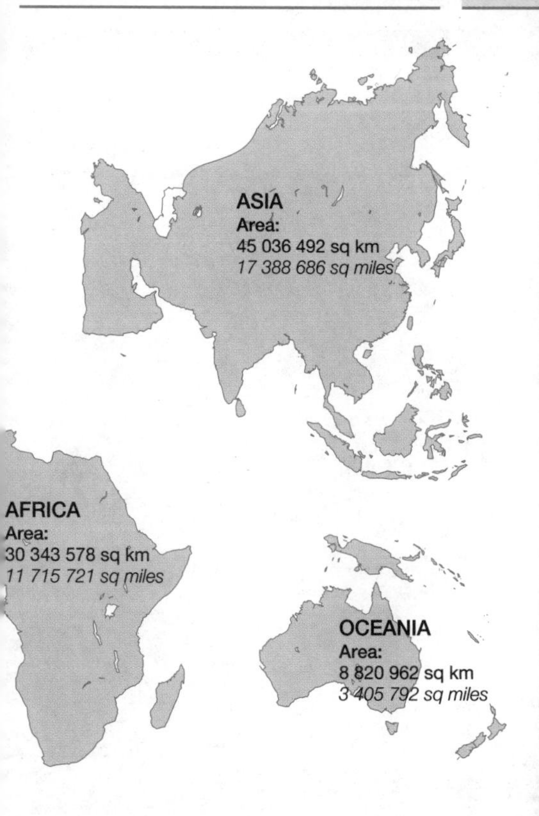
ASIA
Area:
45 036 492 sq km
17 388 686 sq miles
AFRICA
Area:
30 343 578 sq km
11 715 721 sq miles
OCEANIA
Area:
8 820 962 sq km
3 405 792 sq miles

WORLD: PHYSICAL FEATURES

OCEAN
Arctic Circle
Central
Siberia
Siberian
Plateau
West
Siberian
Plain
Ural Mountains
Volga
Lake
Baikal
Sea of
Okhotsk
Bering
Sea
Gobi
ASIA
Aral
Sea
Caspian
Sea
Black Sea
5642
Tien Shan
Kunlun Shan
Himalaya
Mt Everest
8848
Sea
of
Japan
Honshū
East
China
Sea
PACIFIC
Tropic of Cancer
Yangtze
Ganges
Deccan
Indus
Arabian
Peninsula
Red Sea
Nile
Arabian
Sea
Bay
of
Bengal
South
China
Sea
Philippines
Mariana Trench
Challenger
Deep
10920
OCEAN
Micronesia
Ethiopian
Highlands
Maldives
Sri Lanka
Equator
Borneo
Sumatra
Celebes
Puncak Jaya
5030
New
Guinea
Java
Arafura
Sea
Melanesia
Lake
Victoria
Kilimanjaro
5892
Great Rift Valley
Seychelles
INDIAN
Madagascar
Coral
Sea
OCEAN
AUSTRALIA
Great
Victoria
Desert
Great
Australian
Bight
Darling
Great Dividing Ra.
Tropic of Capricorn
Tasman
Sea
Tasmania
New Zealand
Îles Kerguélen
Davis Sea
Antarctic Circle
Ross Sea

WORLD: MOUNTAINS AND VOLCANOES

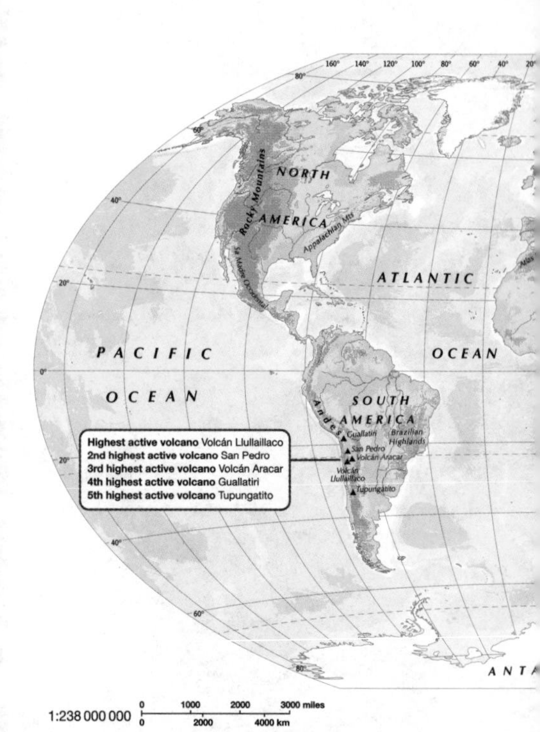

NORTH AMERICA
Rocky Mountains
Appalachian Mts
Sa Madre Occidental
ATLANTIC OCEAN
PACIFIC OCEAN
SOUTH AMERICA
Andes
Brazilian Highlands
Guallatiri
San Pedro
Volcán Aracar
Volcán Llullaillaco
Tupungatito
Highest active volcano Volcán Llullaillaco
2nd highest active volcano San Pedro
3rd highest active volcano Volcán Aracar
4th highest active volcano Guallatiri
5th highest active volcano Tupungatito
1:238 000 000
0 1000 2000 3000 miles
0 2000 4000 km

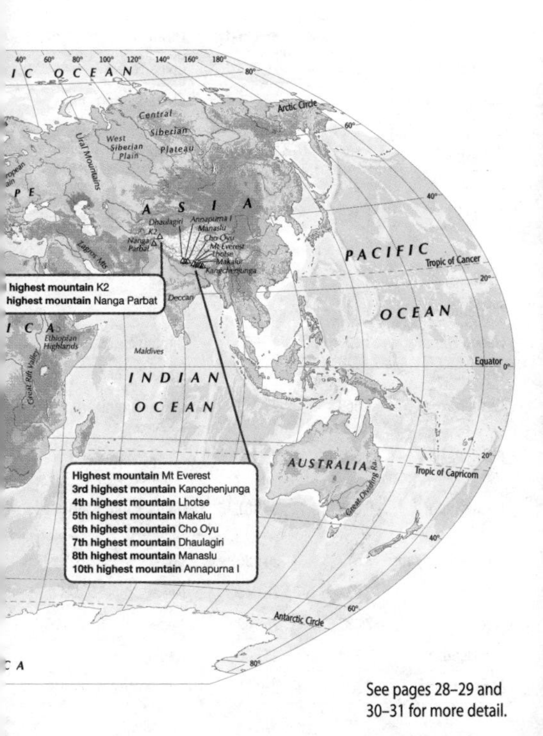

40° 60° 80° 100° 120° 140° 160° 180°
IC OCEAN
80°
Arctic Circle
60°
Central Siberian Plateau
West Siberian Plain
Ural Mountains
PE
ASIA
Dhaulagiri
Annapurna I
Manaslu
K2
Nanga Parbat
Cho Oyu
Mt Everest
Lhotse
Makalu
Kangchenjunga
Zagros Mts
40°
PACIFIC
OCEAN
Tropic of Cancer
20°
highest mountain K2
highest mountain Nanga Parbat
Deccan
ICA
Ethiopian Highlands
Great Rift Valley
Maldives
Equator 0°
INDIAN
OCEAN
AUSTRALIA
Great Dividing Ra.
Tropic of Capricorn
20°
Highest mountain Mt Everest
3rd highest mountain Kangchenjunga
4th highest mountain Lhotse
5th highest mountain Makalu
6th highest mountain Cho Oyu
7th highest mountain Dhaulagiri
8th highest mountain Manaslu
10th highest mountain Annapurna I
40°
60°
Antarctic Circle
80°
CA

See pages 28–29 and 30–31 for more detail.

WORLD: HIGHEST MOUNTAINS

	Height (m)	Height (ft)	Location
Mt Everest	8 848	29 028	China/Nepal
K2	8 611	28 251	China/Jammu and Kashmir
Kangchenjunga	8 586	28 169	India/Nepal
Lhotse	8 516	27 939	China/Nepal
Makalu	8 463	27 765	China/Nepal
Cho Oyu	8 201	26 906	China/Nepal
Dhaulagiri	8 167	26 794	Nepal
Manaslu	8 163	26 781	Nepal
Nanga Parbat	8 126	26 660	Jammu and Kashmir
Annapurna I	8 091	26 545	Nepal
Gasherbrum I	8 068	26 469	China/Jammu and Kashmir
Broad Peak	8 047	26 401	China/Jammu and Kashmir
Gasherbrum II	8 035	26 361	China/Jammu and Kashmir
Xixabangma Feng	8 012	26 286	China
Annapurna II	7 937	26 040	Nepal
Nuptse	7 885	25 869	Nepal
Himalchul	7 864	25 800	Nepal
Masherbrum	7 821	25 659	Jammu and Kashmir

	Height (m)	Height (ft)	Location
Nanda Devi	7 816	25 643	India
Rakaposhi	7 788	25 551	Jammu and Kashmir
Kamet	7 756	25 446	China
Namjagbarwa Feng	7 756	25 446	China
Gurla Mandhata	7 739	25 390	China
Muztag	7 723	25 338	China
Kongur Shan	7 719	25 324	China
Tirich Mir	7 690	25 229	Pakistan
Kula Kangri	7 554	24 783	Bhutan
Muztagata	7 546	24 757	China
Gongga Shan	7 514	24 652	China
Qullai Garmo	7 495	24 590	Tajikistan
Jongsang	7 483	24 550	India/Nepal
Teram Kangri	7 470	24 508	China/Jammu and Kashmir
Pik Pobedy	7 439	24 406	China/Kyrgyzstan
Ganesh I	7 415	24 327	China/Nepal
Churen Himal	7 371	24 183	Nepal
Sad Istragh	7 367	24 170	Afghanistan/Pakistan
Kabru	7 353	24 124	India/Nepal
Chamlang	7 319	24 012	Nepal
Choksiam	7 316	24 002	China
Chomo Lhari	7 313	23 992	Bhutan

WORLD: HIGHEST ACTIVE VOLCANOES

	Height (m)	Height (ft)	Location
World			
Volcán Llullaillaco	6 723	22 057	Chile
San Pedro	6 199	20 338	Chile
Volcán Aracar	6 082	19 954	Argentina
Guallatiri	6 060	19 882	Chile
Tupungatito	6 000	19 685	Chile
Europe			
Mt Etna	3 323	10 902	Italy
Pico	2 351	7 713	Azores
Haakon VII Topp	2 277	7 470	Jan Mayen (Norway)
Hvannadalshnúkur	2 119	6 952	Iceland
Bárđarbunga	2 009	6 591	Iceland
Asia			
Kunlun Shan range	5 808	19 055	China
Mt Ararat	5 165	16 945	Turkey
Sopka Klyuchevskaya	4 750	15 584	Russian Federation
Sopka Ushkovsky	3 943	12 936	Russian Federation
Kūh-e Taftān	4 042	13 261	Iran
Africa			
Meru	4 565	14 977	Tanzania
Mont Cameroun	4 100	13 451	Cameroon
Pico del Teide	3 718	12 198	Canary Islands
Visoke	3 711	12 175	Dem. Rep. Congo/Rwanda
Nyiragongo	3 469	11 381	Dem. Rep. Congo

	Height (m)	Height (ft)	Location
Oceania			
Mt Ruapehu	2 797	9 176	New Zealand
Mt Taranaki	2 518	8 261	New Zealand
Mt Ulawun	2 334	7 657	Papua New Guinea
Mt Bamus	2 249	7 379	Papua New Guinea
Tongariro	1 968	6 457	New Zealand
North America			
Pico de Orizaba	5 610	18 405	Mexico
Volcán Popocatépetl	5 452	17 887	Mexico
Mt Rainier	4 392	14 409	United States of America
Mt Shasta	4 317	14 163	United States of America
Mt Wrangell	4 301	14 111	United States of America
South America			
Volcán Llullaillaco	6 723	22 057	Chile
San Pedro	6 199	20 338	Chile
Volcán Aracar	6 082	19 954	Argentina
Guallatiri	6 060	19 882	Chile
Tupungatito	6 000	19 685	Chile
Antarctica			
Mt Erebus	3 794	12 447	
Mt Melbourne	2 732	8 963	
Buckle Island	1 239	4 065	
Deception Island	576	1 890	

WORLD: RIVERS AND LAKES

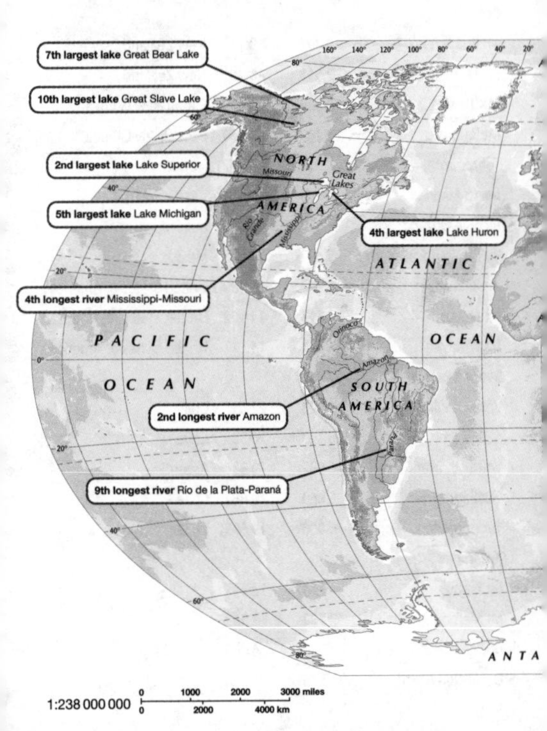
7th largest lake Great Bear Lake
10th largest lake Great Slave Lake
2nd largest lake Lake Superior
5th largest lake Lake Michigan
4th largest lake Lake Huron
4th longest river Mississippi-Missouri
2nd longest river Amazon
9th longest river Río de la Plata-Paraná
NORTH AMERICA
SOUTH AMERICA
PACIFIC OCEAN
ATLANTIC OCEAN
Great Lakes
Missouri
Mississippi
Rio Grande
Orinoco
Amazon
Paraná
ANTA
1:238 000 000
0 1000 2000 3000 miles
0 2000 4000 km

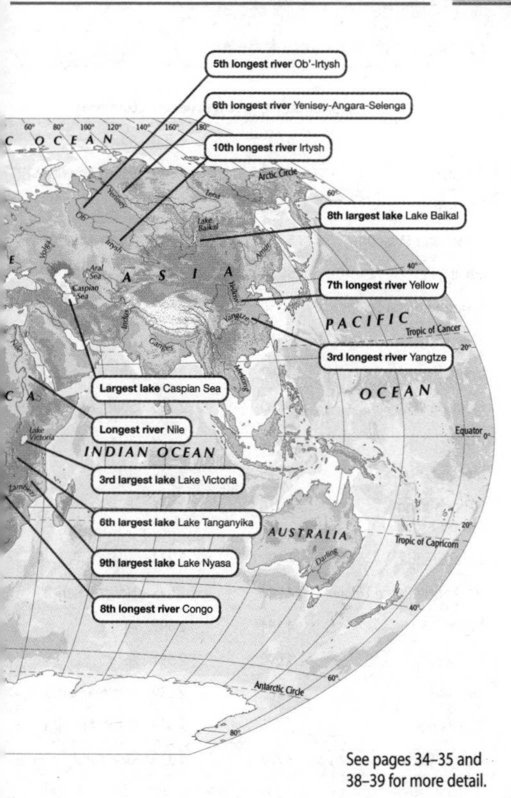
5th longest river Ob'-Irtysh
6th longest river Yenisey-Angara-Selenga
10th longest river Irtysh
8th largest lake Lake Baikal
7th longest river Yellow
3rd longest river Yangtze
Largest lake Caspian Sea
Longest river Nile
3rd largest lake Lake Victoria
6th largest lake Lake Tanganyika
9th largest lake Lake Nyasa
8th longest river Congo
OCEAN
Arctic Circle
Yenisey
Ob'
Lena
Lake Baikal
Irtysh
Volga
Amur
Aral Sea
Caspian Sea
ASIA
Yellow
Yangtze
Indus
Ganges
Mekong
Nile
PACIFIC
OCEAN
Tropic of Cancer
Equator
Lake Victoria
INDIAN OCEAN
Zambezi
AUSTRALIA
Darling
Tropic of Capricorn
Antarctic Circle

See pages 34–35 and 38–39 for more detail.

WORLD: LONGEST RIVERS

	Length (km)	Length (miles)	Continent
Nile	6 695	4 160	Africa
Amazon	6 516	4 049	South America
Yangtze	6 380	3 964	Asia
Mississippi-Missouri	5 969	3 709	North America
Ob'-Irtysh	5 568	3 459	Asia
Yenisey-Angara-Selenga	5 550	3 448	Asia
Yellow	5 464	3 395	Asia
Congo	4 667	2 900	Africa
Río de la Plata-Paraná	4 500	2 796	South America
Irtysh	4 440	2 759	Asia
Mekong	4 425	2 749	Asia
Heilong Jiang (Amur)-Argun'	4 416	2 744	Asia
Lena-Kirenga	4 400	2 734	Asia
Mackenzie-Peace-Finlay	4 241	2 635	North America
Niger	4 184	2 599	Africa
Yenisey	4 090	2 541	Asia
Missouri	4 086	2 539	North America
Mississippi	3 765	2 339	North America
Murray-Darling	3 750	2 330	Oceania

	Length (km)	Length (miles)	Continent
Ob'	3 701	2 300	Asia
Volga	3 688	2 291	Europe
Purus	3 218	2 000	South America
Madeira	3 200	1 988	South America
Yukon	3 185	1 979	North America
Indus	3 180	1 976	Asia
Syrdar'ya	3 078	1 913	Asia
St Lawrence	3 058	1 900	North America
Rio Grande	3 057	1 899	North America
São Francisco	2 900	1 802	South America
Danube	2 850	1 770	Europe
Brahmaputra	2 840	1 765	Asia
Salween	2 816	1 750	Asia
Euphrates	2 815	1 749	Asia
Tarim He	2 750	1 708	Asia
Tocantins	2 750	1 708	South America
Darling	2 739	1 702	Oceania
Zambezi	2 736	1 700	Africa
Araguaia	2 627	1 632	South America
Paraguay	2 600	1 615	South America
Murray	2 589	1 608	Oceania

WORLD: DRAINAGE BASINS, WATERFALLS AND CANYONS

WORLD'S LARGEST DRAINAGE BASINS

	Area (sq km)	Area (sq miles)	Continent
Amazon	7 050 000	2 722 000	South America
Congo	3 700 000	1 429 000	Africa
Nile	3 349 000	1 293 000	Africa
Mississippi-Missouri	3 250 000	1 255 000	North America
Río de la Plata-Paraná	3 100 000	1 197 000	South America
Ob'-Irtysh	2 990 000	1 154 000	Asia
Yenisey-Angara-Selenga	2 580 000	996 000	Asia
Lena-Kirenga	2 490 000	961 000	Asia
Yangtze	1 959 000	756 000	Asia
Niger	1 890 000	730 000	Africa
Heilong Jiang (Amur)-Argun'	1 855 000	716 000	Asia
Mackenzie-Peace-Finlay	1 805 000	697 000	North America
Ganges-Brahmaputra	1 621 000	626 000	Asia
St Lawrence-St Louis	1 463 000	565 000	North America
Volga	1 380 000	533 000	Europe
Zambezi	1 330 000	514 000	Africa
Indus	1 166 000	450 000	Asia
Nelson-Saskatchewan	1 150 000	444 000	North America
Shaţţ al'Arab	1 114 000	430 000	Asia
Murray-Darling	1 058 000	408 000	Oceania

WORLD'S HIGHEST WATERFALLS

	Height (m)	Height (ft)	Location
Angel Falls	979	3 212	Venezuela
Tugela	948	3 110	South Africa
Utigård	800	2 625	Norway
Mongefossen	774	2 539	Norway
Mtarazi	762	2 500	Zimbabwe
Yosemite	739	2 425	USA
Mardalsfossen	657	2 155	Norway
Tyssestrengane	646	2 119	Norway
Cuquenan	610	2 001	Venezuela
Sutherland	580	1 903	New Zealand

MAJOR RIVER CANYONS

	Height (m)	Height (ft)	Location
Yarlung Zangbo Canyon	5 382	17 657	China (Tibet)
Kali Gandaki Canyon	4 403	14 445	Nepal
Cotahuasi Canyon	3 535	11 598	Peru
Colca Canyon	3 200	10 499	Peru
Hells Canyon	2 412	7 913	USA
Urique Canyon	1 879	6 165	Mexico
Grand Canyon	1 829	6 000	USA

WORLD: LARGEST AND DEEPEST LAKES

LARGEST LAKES

	Area (sq km)	Area (sq miles)	Continent
Caspian Sea	371 000	143 243	Asia/Europe
Lake Superior	82 100	31 698	North America
Lake Victoria	68 800	26 563	Africa
Lake Huron	59 600	23 011	North America
Lake Michigan	57 800	22 316	North America
Lake Tanganyika	32 900	12 702	Africa
Great Bear Lake	31 328	12 095	North America
Lake Baikal	30 500	11 776	Asia
Lake Nyasa	30 044	11 600	Africa
Great Slave Lake	28 568	11 030	North America
Lake Erie	25 700	9 922	North America
Lake Winnipeg	24 387	9 415	North America
Lake Ontario	18 960	7 320	North America
Lake Ladoga	18 390	7 100	Europe
Lake Balkhash	17 400	6 718	Asia
Aral Sea	17 158	6 625	Asia
Lake Onega	9 600	3 706	Europe
Lake Volta	8 485	3 276	Africa
Lake Titicaca	8 340	3 220	South America
Lake Nicaragua	8 150	3 147	North America

DEEPEST LAKES			
	Depth (m)	Depth (ft)	Continent
World			
Lake Baikal	1 741	5 712	Asia
Lake Tanganyika	1 471	4 826	Africa
Caspian Sea	1 025	3 363	Europe/Asia
Lake Nyasa	706	2 316	Africa
Ysyk-Köl	702	2 303	Asia
Europe			
Hornindal Lake	514	1 686	Norway
Asia			
Lake Baikal	1 741	5 712	Russian Federation
Africa			
Lake Tanganyika	1 471	4 826	Burundi/Dem. Rep. Congo/Tanzania/ Zambia
Oceania			
Lake Manapouri	443	1 453	New Zealand
North America			
Great Slave Lake	614	2 014	Canada
South America			
Lago General Carrera	596	1 955	Chile

WORLD: ISLANDS AND DESERTS

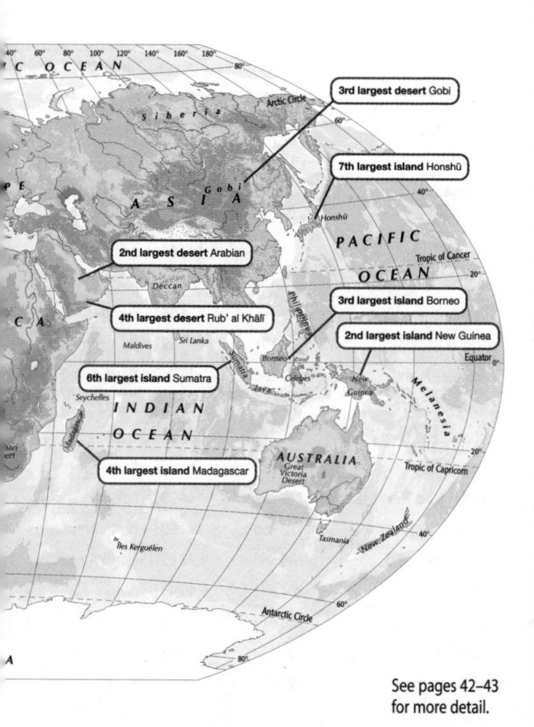

See pages 42–43 for more detail.

WORLD: LARGEST ISLANDS AND DESERTS

LARGEST ISLANDS

	Area (sq km)	Area (sq miles)	Continent
Greenland	2 175 600	840 004	North America
New Guinea	808 510	312 167	Oceania
Borneo	745 561	287 863	Asia
Madagascar	587 040	266 657	Africa
Baffin Island	507 451	195 927	North America
Sumatra	473 606	182 860	Asia
Honshū	227 414	87 805	Asia
Great Britain	218 476	84 354	Europe
Victoria Island	217 291	83 897	North America
Ellesmere Island	196 236	75 767	North America
Celebes	189 216	73 057	Asia
South Island, New Zealand	151 215	58 384	Oceania
Java	132 188	51 038	Asia
North Island, New Zealand	115 777	44 702	Oceania
Cuba	110 860	42 803	North America
Newfoundland	108 860	42 031	North America
Luzon	104 690	40 421	Asia
Iceland	102 820	39 699	Europe
Mindanao	94 630	36 537	Asia
Novaya Zemlya	90 650	35 000	Europe

LARGEST DESERTS

	Area (sq km)	Area (sq miles)	Continent
Sahara	9 065 000	3 500 000	Africa
Arabian	2 300 000	888 000	Asia
Gobi	1 300 000	502 000	Asia
Rub' al Khālī	750 000	290 000	Asia
Patagonian	673 000	260 000	South America
Great Victoria	647 000	250 000	Oceania
Kalahari	582 000	225 000	Africa
Great Basin	492 000	190 000	North America
Chihuahuan	453 000	175 000	North America
Great Sandy	338 000	131 000	Oceania
Colorado	337 000	130 000	North America
Takla Makan	324 000	125 000	Asia
Gibson	311 000	120 000	Oceania
Sonoran	311 000	120 000	North America
Kara Kum	310 000	120 000	Asia
Iranian	260 000	100 000	Asia
Kyzyl Kum	260 000	100 000	Asia
Somali	260 000	100 000	Africa
Syrian	260 000	100 000	Asia
Thar	260 000	100 000	Asia

EUROPE: PHYSICAL FEATURES

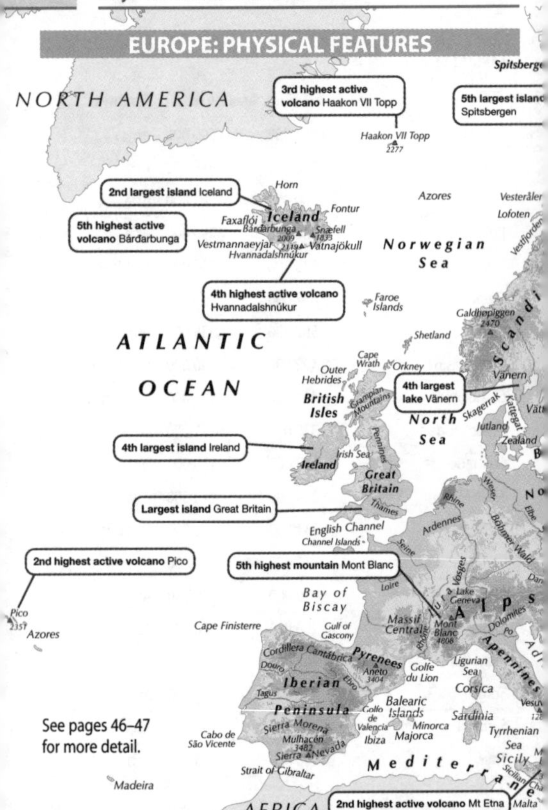

3rd highest active volcano Haakon VII Topp
5th largest island Spitsbergen
2nd largest island Iceland
5th highest active volcano Bárdarbunga
4th highest active volcano Hvannadalshnúkur
4th largest lake Vänern
4th largest island Ireland
Largest island Great Britain
2nd highest active volcano Pico
5th highest mountain Mont Blanc
2nd highest active volcano Mt Etna
NORTH AMERICA
Haakon VII Topp
2277
Horn
Iceland
Fontur
Faxaflói
Bárdarbunga
2009
Snæfell
1833
Vestmannaeyjar
2119
Vatnajökull
Hvannadalshnúkur
Azores
Vesterålen
Lofoten
Vestfjorden
Norwegian Sea
Faroe Islands
Galdhøpiggen
2470
ATLANTIC OCEAN
Shetland
Cape Wrath
Orkney
Outer Hebrides
British Isles
Grampian Mountains
Vänern
North Sea
Skagerrak
Kattegat
Jutland
Zealand
Pennines
Irish Sea
Ireland
Great Britain
Weser
Rhine
Elbe
Thames
Ardennes
Böhmer Wald
English Channel
Channel Islands
Seine
Vosges
Loire
Lake Geneva
Alps
Dolomites
Bay of Biscay
Jura
Pico
2351
Azores
Cape Finisterre
Gulf of Gascony
Massif Central
Rhône
Mont Blanc
4808
Po
Cordillera Cantábrica
Pyrenees
Aneto
3404
Apennines
Douro
Golfe du Lion
Ligurian Sea
Iberian Peninsula
Ebro
Corsica
Tagus
Balearic Islands
Golfo de Valencia
Sardinia
Sierra Morena
Minorca
Cabo de São Vicente
Mulhacén
3482
Sierra Nevada
Ibiza
Majorca
Tyrrhenian Sea
Sicily
Strait of Gibraltar
Mediterranean
Madeira
AFRICA
Malta

See pages 46–47 for more detail.

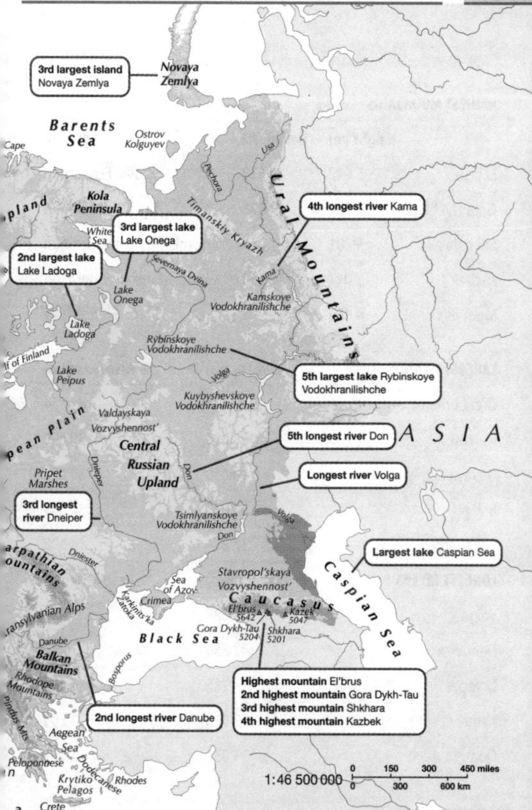
3rd largest island Novaya Zemlya
Novaya Zemlya
Barents Sea
Ostrov Kolguyev
Cape
Kola Peninsula
White Sea
3rd largest lake Lake Onega
2nd largest lake Lake Ladoga
Lake Onega
Lake Ladoga
Pechora
Usa
Timanskiy Kryazh
Severnaya Dvina
Ural Mountains
4th longest river Kama
Kama
Kamskoye Vodokhranilishche
Rybinskoye Vodokhranilishche
5th largest lake Rybinskoye Vodokhranilishche
Lake Peipus
Volga
Kuybyshevskoye Vodokhranilishche
Valdayskaya Vozvyshennost'
ASIA
5th longest river Don
Central Russian Upland
Don
Dnieper
Pripet Marshes
Longest river Volga
3rd longest river Dneiper
Tsimlyanskoye Vodokhranilishche
Don
Volga
Dniester
Largest lake Caspian Sea
Caspian Sea
Sea of Azov
Stavropol'skaya Vozvyshennost'
Caucasus
El'brus 5642
Kazbek 5047
Crimea
Karkinits'ka Zatoka
Transylvanian Alps
Gora Dykh-Tau 5204
Shkhara 5201
Black Sea
Danube
Balkan Mountains
Rhodope Mountains
Bosporus
Pindus Mts
Highest mountain El'brus
2nd highest mountain Gora Dykh-Tau
3rd highest mountain Shkhara
4th highest mountain Kazbek
2nd longest river Danube
Aegean Sea
Peloponnese
Dodecanese
Rhodes
Krytiko Pelagos
Crete
1:46 500 000
0 150 300 450 miles
0 300 600 km

EUROPE: PHYSICAL EXTREMES

HIGHEST MOUNTAINS			
	Height (m)	Height (ft)	Location
El'brus	5 642	18 510	Russian Federation
Gora Dykh-Tau	5 204	17 073	Russian Federation
Shkhara	5 201	17 063	Georgia/Russian Federation
Kazbek	5 047	16 558	Georgia/Russian Federation
Mont Blanc	4 808	15 774	France/Italy

LARGEST ISLANDS	Area (sq km)	Area (sq miles)
Great Britain	218 476	84 354
Iceland	102 820	39 699
Novaya Zemlya	90 650	35 000
Ireland	83 045	32 064
Spitzbergen	37 814	14 600

LONGEST RIVERS	Length (km)	Length (miles)
Volga	3 688	2 291
Danube	2 850	1 770
Dnieper	2 285	1 419
Kama	2 028	1 260
Don	1 931	1 199

LARGEST LAKES	Area (sq km)	Area (sq miles)
Caspian Sea	371 000	143 243
Lake Ladoga	18 390	7 100
Lake Onega	9 600	3 706
Vänern	5 585	2 156
Rybinskoye Vodokhranilishche	5 180	2 000

HIGHEST ACTIVE VOLCANOES

	Height (m)	Height (ft)	Location
Mt Etna	3 323	10 902	Italy
Pico	2 351	7 713	Azores
Haakon VII Topp	2 277	7 470	Jan Mayen (Norway)
Hvannadalshnúkur	2 119	6 952	Iceland
Bárðarbunga	2 009	6 591	Iceland

LAND AREA

Total land area 9 908 599 sq km / 3 825 731 sq miles

Most northerly point	Ostrov Rudol'fa, Russian Federation
Most southerly point	Gavdos, Crete, Greece
Most westerly point	Bjargtangar, Iceland
Most easterly point	Mys Flissingskiy, Russian Federation
Lowest point	Caspian Sea, Asia/Europe

ASIA: PHYSICAL FEATURES

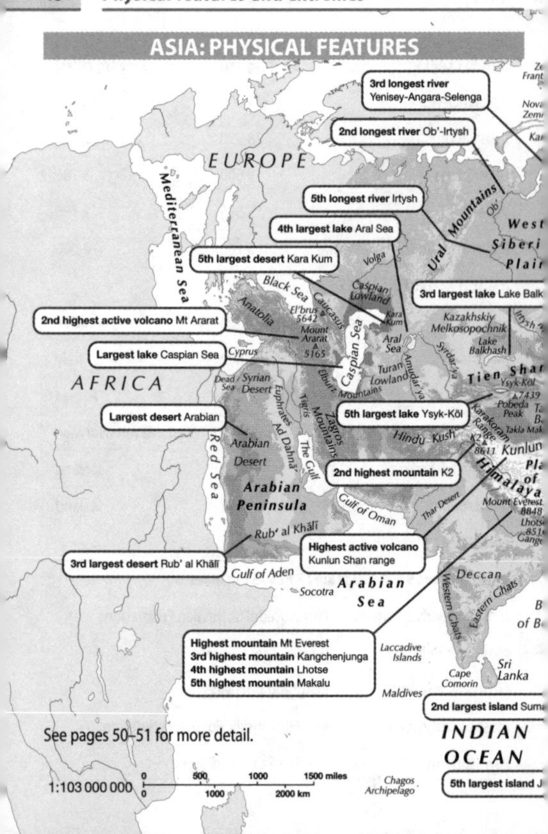
3rd longest river Yenisey-Angara-Selenga
2nd longest river Ob'-Irtysh
5th longest river Irtysh
4th largest lake Aral Sea
5th largest desert Kara Kum
3rd largest lake Lake Balk
2nd highest active volcano Mt Ararat
Largest lake Caspian Sea
Largest desert Arabian
5th largest lake Ysyk-Köl
2nd highest mountain K2
Highest active volcano Kunlun Shan range
3rd largest desert Rub' al Khālī
Highest mountain Mt Everest
3rd highest mountain Kangchenjunga
4th highest mountain Lhotse
5th highest mountain Makalu
2nd largest island Suma
5th largest island J
EUROPE
AFRICA
Mediterranean Sea
Black Sea
Caspian Sea
Ural Mountains
West Siberi Plair
Volga
Caspian Lowland
Caucasus
El'brus 5642
Mount Ararat 5165
Anatolia
Cyprus
Kara Kum
Aral Sea
Kazakhskiy Melkosopochnik
Lake Balkhash
Syrdar'ya
Amudar'ya
Turan Lowland
Elburz Mountains
Dead Sea
Syrian Desert
Euphrates
Tigris
Zagros Mountains
Tien Shan
Ysyk-Köl
7439 Pobeda Peak
Karakoram Range
Hindu Kush
K2 8611
Kunlun
Himalaya
Mount Everest 8848
Red Sea
Arabian Desert
Ad Dahnā'
The Gulf
Arabian Peninsula
Gulf of Oman
Thar Desert
Rub' al Khālī
Gulf of Aden
Socotra
Arabian Sea
Deccan
Western Ghats
Eastern Ghats
Laccadive Islands
Maldives
Cape Comorin
Sri Lanka
INDIAN OCEAN
Chagos Archipelago
1:103 000 000
0 500 1000 1500 miles
0 1000 2000 km

See pages 50–51 for more detail.

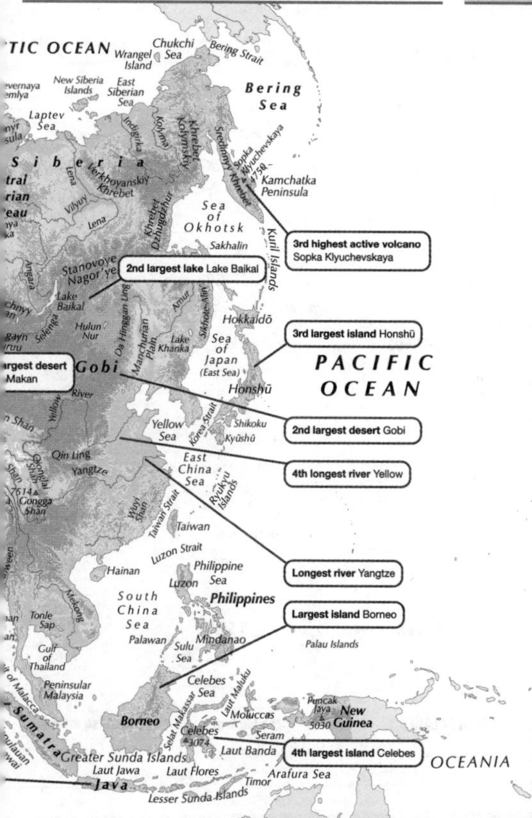

3rd highest active volcano Sopka Klyuchevskaya
2nd largest lake Lake Baikal
3rd largest island Honshū
largest desert Makan
2nd largest desert Gobi
4th longest river Yellow
Longest river Yangtze
Largest island Borneo
4th largest island Celebes
PACIFIC OCEAN
Bering Sea
Sea of Okhotsk
Sea of Japan (East Sea)
Yellow Sea
East China Sea
South China Sea
Philippine Sea
Celebes Sea
Sulu Sea
Arafura Sea
OCEANIA
Siberia
Kamchatka Peninsula
Sopka Klyuchevskaya 4750
Bering Strait
Chukchi Sea
Wrangel Island
New Siberia Islands
East Siberian Sea
Laptev Sea
Sakhalin
Kuril Islands
Hokkaidō
Honshū
Shikoku
Kyūshū
Lake Baikal
Hulun Nur
Lake Khanka
Gobi
Qin Ling
Yangtze
Taiwan
Taiwan Strait
Luzon Strait
Hainan
Luzon
Philippines
Mindanao
Palawan
Palau Islands
Borneo
Celebes 3074
Moluccas
Seram
Puncak Jaya 5030
New Guinea
Laut Banda
Laut Jawa
Laut Flores
Timor
Java
Lesser Sunda Islands
Greater Sunda Islands
Sumatra
Peninsular Malaysia
Gulf of Thailand
Tonle Sap
Mekong
Gongga Shan 7514
Ryukyu Islands
Korea Strait
Manchurian Plain
Da Hinggan Ling
Sikhote-Alin
Amur
Stanovoye Nagor'ye
Khrebet Dzhugdzhur
Verkhoyanskiy Khrebet
Lena
Vilyuy
Angara
Selenga
Indigirka
Kolyma
Khrebet Kolymskiy
Sredinnyy Khrebet
Wuyi Shan
Laut Maluku
Selat Makassar

ASIA: PHYSICAL EXTREMES

HIGHEST MOUNTAINS			
	Height (m)	Height (ft)	Location
Mt Everest	8 848	29 028	China/Nepal
K2	8 611	28 251	China/Jammu and Kashmir
Kangchenjunga	8 586	28 169	India/Nepal
Lhotse	8 516	27 939	China/Nepal
Makalu	8 463	27 765	China/Nepal

LARGEST ISLANDS	Area (sq km)	Area (sq miles)
Borneo	745 561	287 863
Sumatra	473 606	182 860
Honshū	227 414	87 805
Celebes	189 216	73 057
Java	132 188	51 038

LONGEST RIVERS	Length (km)	Length (miles)
Yangtze	6 380	3 964
Ob'-Irtysh	5 568	3 459
Yenisey-Angara-Selenga	5 550	3 448
Yellow	5 464	3 395
Irtysh	4 440	2 759

LARGEST LAKES	Area (sq km)	Area (sq miles)
Caspian Sea	371 000	143 243

Lake Baikal		30 500	11 776
Lake Balkhash		17 400	6 718
Aral Sea		17 158	6 625
Ysyk-Köl		6 200	2 393
HIGHEST ACTIVE VOLCANOES			
	Height (m)	**Height (ft)**	**Location**
Kunlun Shan range	5 808	19 055	China
Mt Ararat	5 165	16 945	Turkey
Sopka Klyuchevskaya	4 750	15 584	Russian Federation
LARGEST DESERTS		**Area (sq km)**	**Area (sq miles)**
Arabian		2 300 000	888 000
Gobi		1 300 000	502 000
Rub' al Khālī		750 000	290 000
Takla Makan		324 000	125 000
Kara Kum		310 000	120 000
LAND AREA			
Total land area 45 036 492 sq km/ 17 388 686 sq miles			
Most northerly point	Mys Arkticheskiy, Russian Federation		
Most southerly point	Pamana, Indonesia		
Most westerly point	Bozcaada, Turkey		
Most easterly point	Mys Dezhneva, Russian Federation		
Lowest point	Dead Sea		

AFRICA: PHYSICAL FEATURES

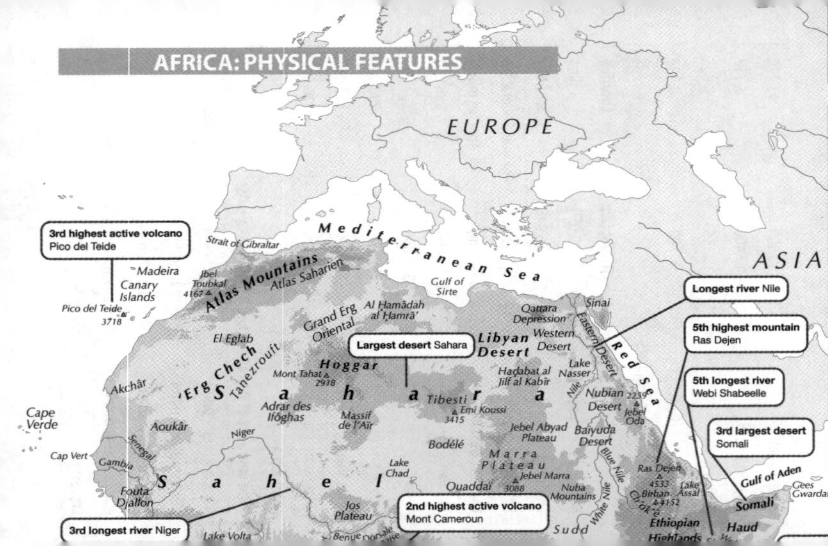

3rd highest active volcano Pico del Teide

Longest river Nile

5th highest mountain Ras Dejen

5th longest river Webi Shabeelle

3rd largest desert Somali

Largest desert Sahara

2nd highest active volcano Mont Cameroun

3rd longest river Niger

EUROPE
ASIA
Mediterranean Sea
Strait of Gibraltar
Madeira
Canary Islands
Pico del Teide 3718
Jbel Toubkal 4167
Atlas Mountains
Atlas Saharien
Gulf of Sirte
Grand Erg Oriental
Al Hamādah al Hamrā'
El Eglab
'Erg Chech
Tanezrouft
Hoggar
Mont Tahat 2918
Libyan Desert
Qattara Depression
Western Desert
Sinai
Eastern Desert
Red Sea
Lake Nasser
Hadabat al Jilf al Kabīr
Nile
Nubian Desert
Jebel Oda 2259
Akchâr
Sahara
Tibesti
Emi Koussi 3415
Adrar des Ifôghas
Massif de l'Aïr
Cape Verde
Aoukâr
Niger
Cap Vert
Senegal
Gambia
Bodélé
Jebel Abyad Plateau
Baiyuda Desert
Marra Plateau
Jebel Marra 3088
Lake Chad
Blue Nile
Ras Dejen 4533
Birhan 4152
Lake Assal
Gulf of Aden
Gees Gwardafuy
Sahel
Fouta Djallon
Ouaddaï
Nuba Mountains
White Nile
Ch'ok'e
Somali
Jos Plateau
Sudd
Ethiopian Highlands
Haud
Lake Volta
Benue
Dorsale

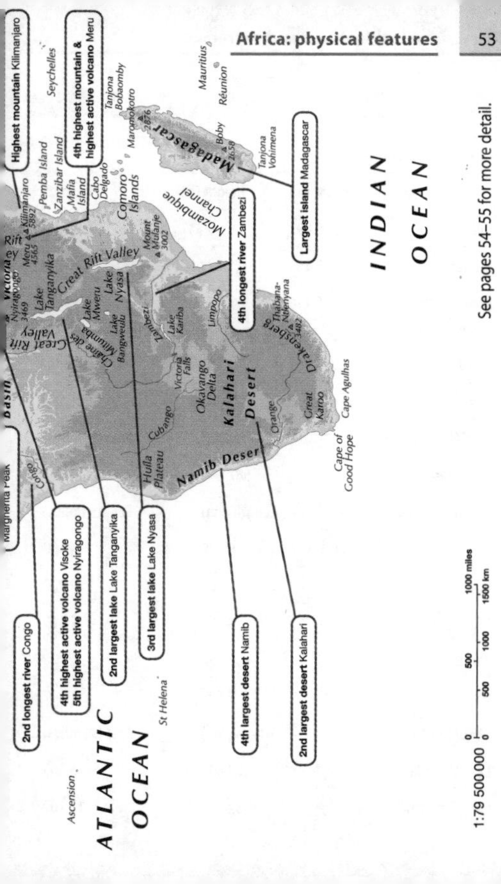

2nd longest river Congo

4th highest active volcano Visoke
5th highest active volcano Nyiragongo

2nd largest lake Lake Tanganyika

3rd largest lake Lake Nyasa

4th largest desert Namib

2nd largest desert Kalahari

Highest mountain Kilimanjaro

4th highest mountain & highest active volcano Meru

4th longest river Zambezi

Largest island Madagascar

ATLANTIC OCEAN

INDIAN OCEAN

Ascension
St Helena
Margherita Peak
Basin
Congo
Victoria
Nyiragongo 3469
Great Rift Valley
Lake Tanganyika
Meru 4565
Kilimanjaro 5892
Rift
Pemba Island
Zanzibar Island
Mafia Island
Seychelles
Chaîne des Mitumba
Lake Mweru
Lake Bangweulu
Lake Nyasa
Great Rift Valley
Cabo Delgado
Comoro Islands
Tanjona Bobaomby
Maromokotro 2876
Huíla Plateau
Cubango
Zambezi
Mount Mulanje 3002
Victoria Falls
Lake Kariba
Mozambique Channel
Madagascar
Okavango Delta
Limpopo
Kalahari Desert
Namib Desert
Boby 2658
Mauritius
Réunion
Tanjona Vohimena
Orange
Thabana-Ntlenyana 3482
Drakensberg
Great Karoo
Cape of Good Hope
Cape Agulhas

1:79 500 000
0 500 1000 miles
0 500 1000 1500 km

See pages 54–55 for more detail.

AFRICA: PHYSICAL EXTREMES

HIGHEST MOUNTAINS			
	Height (m)	Height (ft)	Location
Kilimanjaro	5 892	19 331	Tanzania
Mt Kenya	5 199	17 057	Kenya
Margherita Peak	5 110	16 765	Dem. Rep. Congo/Uganda
Meru	4 565	14 977	Tanzania
Ras Dejen	4 533	14 872	Ethiopia

LARGEST ISLANDS	Area (sq km)	Area (sq miles)
Madagascar	587 040	226 657

LONGEST RIVERS	Length (km)	Length (miles)
Nile	6 695	4 160
Congo	4 667	2 900
Niger	4 184	2 599
Zambezi	2 736	1 700
Webi Shabeelle	2 490	1 547

LARGEST LAKES	Area (sq km)	Area (sq miles)
Lake Victoria	68 800	26 563
Lake Tanganyika	32 900	12 702

Lake Nyasa		30 044	11 600
Lake Volta		8 485	3 276
Lake Turkana		6 475	2 500
HIGHEST ACTIVE VOLCANOES			
	Height (m)	**Height (ft)**	**Location**
Meru	4 565	14 977	Tanzania
Mont Cameroun	4 100	13 451	Cameroon
Pico del Teide	3 718	12 198	Canary Islands
Visoke	3 711	12 175	Dem. Rep. Congo/Rwanda
Nyiragongo	3 469	11 381	Dem. Rep. Congo
LARGEST DESERTS		**Area (sq km)**	**Area (sq miles)**
Sahara		9 065 000	3 500 000
Kalahari		582 000	225 000
Somali		260 000	100 000
Namib		135 000	52 000

LAND AREA

Total land area 30 343 578 sq km / 11 715 721 sq miles

Most northerly point	La Galite, Tunisia
Most southerly point	Cape Agulhas, South Africa
Most westerly point	Santo Antao, Cape Verde
Most easterly point	Raas Xaafuun, Somalia
Lowest point	Lake Assal, Djibouti

OCEANIA: PHYSICAL FEATURES

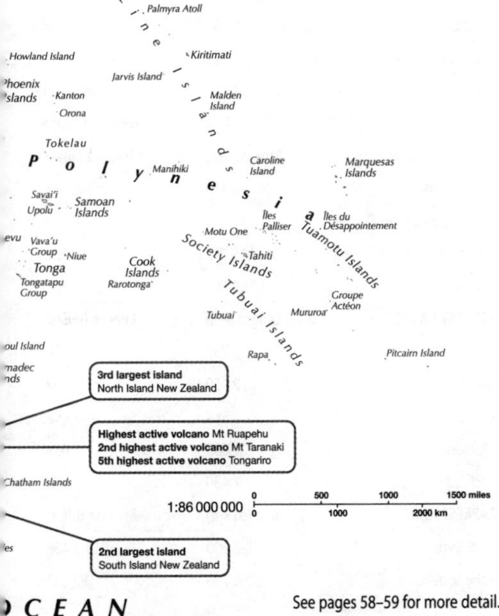

See pages 58–59 for more detail.

OCEANIA: PHYSICAL EXTREMES

HIGHEST MOUNTAINS			
	Height (m)	Height (ft)	Location
Puncak Jaya	5 030	16 502	Indonesia
Puncak Trikora	4 730	15 518	Indonesia
Puncak Mandala	4 700	15 420	Indonesia
Puncak Yamin	4 595	15 075	Indonesia
Mt Wilhelm	4 509	14 793	Papua New Guinea

LARGEST ISLANDS	Area (sq km)	Area (sq miles)
New Guinea	808 510	312 167
South Island, New Zealand	151 215	58 384
North Island, New Zealand	115 777	44 702
Tasmania	67 800	26 178

LONGEST RIVERS	Length (km)	Length (miles)
Murray-Darling	3 750	2 330
Darling	2 739	1 702
Murray	2 589	1 608
Murrumbidgee	1 690	1 050
Lachlan	1 480	919

LARGEST LAKES	Area (sq km)	Area (sq miles)
Lake Eyre	0-8 900	0-3 436
Lake Torrens	0-5 780	0-2 232

HIGHEST ACTIVE VOLCANOES			
	Height (m)	Height (ft)	Location
Mt Ruapehu	2 797	9 176	New Zealand
Mt Taranaki	2 518	8 261	New Zealand
Mt Ulawun	2 334	7 657	Papua New Guinea
Mt Bamus	2 249	7 379	Papua New Guinea
Tongariro	1 968	6 457	New Zealand

LARGEST DESERTS	Area (sq km)	Area (sq miles)
Great Victoria	647 000	250 000
Great Sandy	338 000	131 000
Gibson	311 000	120 000
Simpson	145 000	56 000

LAND AREA	
Total land area 8 844 516 sq km / 3 414 887 sq miles	
(includes New Guinea and Pacific Island nations)	
Most northerly point	Eastern Island, North Pacific Ocean
Most southerly point	Macquarie Island, South Pacific Ocean
Most westerly point	Cape Inscription, Australia
Most easterly point	Île Clipperton, North Pacific Ocean
Lowest point	Lake Eyre, Australia

NORTH AMERICA: PHYSICAL FEATURES

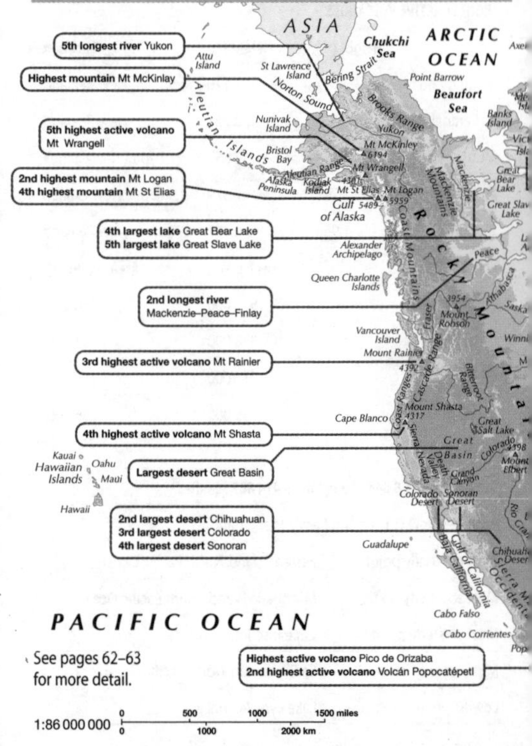

See pages 62–63 for more detail.

1:86 000 000 — 0, 500, 1000, 1500 miles; 0, 1000, 2000 km

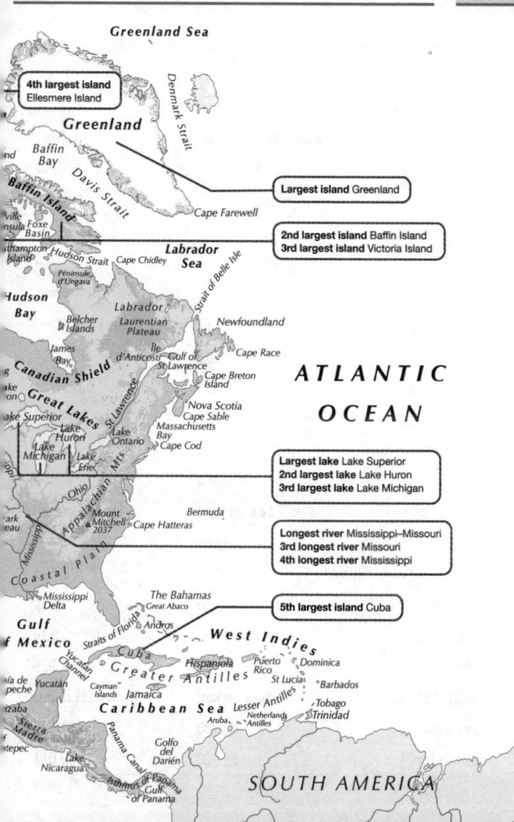
Greenland Sea
4th largest island Ellesmere Island
Greenland
Denmark Strait
Baffin Bay
Davis Strait
Baffin Island
Largest island Greenland
Cape Farewell
Foxe Basin
2nd largest island Baffin Island
3rd largest island Victoria Island
Hudson Strait
Cape Chidley
Labrador Sea
Strait of Belle Isle
Péninsule d'Ungava
Hudson Bay
Labrador
Belcher Islands
Laurentian Plateau
Newfoundland
James Bay
Île d'Anticosti
Gulf of St Lawrence
Cape Race
Canadian Shield
Cape Breton Island
ATLANTIC OCEAN
Great Lakes
St Lawrence
Nova Scotia
Cape Sable
Lake Superior
Lake Huron
Lake Ontario
Massachusetts Bay
Cape Cod
Lake Michigan
Lake Erie
Largest lake Lake Superior
2nd largest lake Lake Huron
3rd largest lake Lake Michigan
Ohio
Appalachian Mts
Mount Mitchell 2037
Bermuda
Cape Hatteras
Mississippi
Longest river Mississippi–Missouri
3rd longest river Missouri
4th longest river Mississippi
Coastal Plain
Mississippi Delta
The Bahamas
Great Abaco
5th largest island Cuba
Andros
Gulf of Mexico
Straits of Florida
West Indies
Yucatán Channel
Cuba
Hispaniola
Puerto Rico
Dominica
Greater Antilles
St Lucia
Barbados
Yucatán
Cayman Islands
Jamaica
Lesser Antilles
Tobago
Trinidad
Caribbean Sea
Aruba
Netherlands Antilles
Sierra Madre
Panama Canal
Golfo del Darién
Lake Nicaragua
Isthmus of Panama
Gulf of Panama
SOUTH AMERICA

NORTH AMERICA: PHYSICAL EXTREMES

HIGHEST MOUNTAINS	Height (m)	Height (ft)	Location
Mt McKinley	6 194	20 321	USA
Mt Logan	5 959	19 550	Canada
Pico de Orizaba	5 610	18 855	Mexico
Mt St Elias	5 489	18 008	USA
Volcán Popocatépetl	5 452	17 887	Mexico

LARGEST ISLANDS	Area (sq km)	Area (sq miles)
Greenland	2 175 600	840 004
Baffin Island	507 451	195 927
Victoria Island	217 291	83 897
Ellesmere Island	196 236	75 767
Cuba	110 860	42 803

LONGEST RIVERS	Length (km)	Length (miles)
Mississippi-Missouri	5 969	3 709
Mackenzie-Peace-Finlay	4 241	2 635
Missouri	4 086	2 539
Mississippi	3 765	2 339
Yukon	3 185	1 979

LARGEST LAKES	Area (sq km)	Area (sq miles)
Lake Superior	82 100	31 698
Lake Huron	59 600	23 011

Lake Michigan		57 800	22 316
Great Bear Lake		31 328	12 095
Great Slave Lake		28 568	11 030
HIGHEST ACTIVE VOLCANOES			
	Height (m)	**Height (ft)**	**Location**
Pico de Orizaba	5 610	18 405	Mexico
Volcán Popocatépetl	5 452	17 887	Mexico
Mt Rainier	4 392	14 409	USA
Mt Shasta	4 317	14 163	USA
Mt Wrangell	4 301	14 111	USA
LARGEST DESERTS		**Area (sq km)**	**Area (sq miles)**
Great Basin		492 000	190 000
Chihuahuan		453 000	175 000
Colorado		337 000	130 000
Sonoran		311 000	120 000

LAND AREA	
Total land area 24 680 331 sq km / 9 529 129 sq miles	
(including Hawaiian Islands)	
Most northerly point	Kap Morris Jesup, Greenland
Most southerly point	Punta Mariato, Panama
Most westerly point	Attu Island, Aleutian Islands, USA
Most easterly point	Nordøstrundingen, Greenland
Lowest point	Death Valley, USA

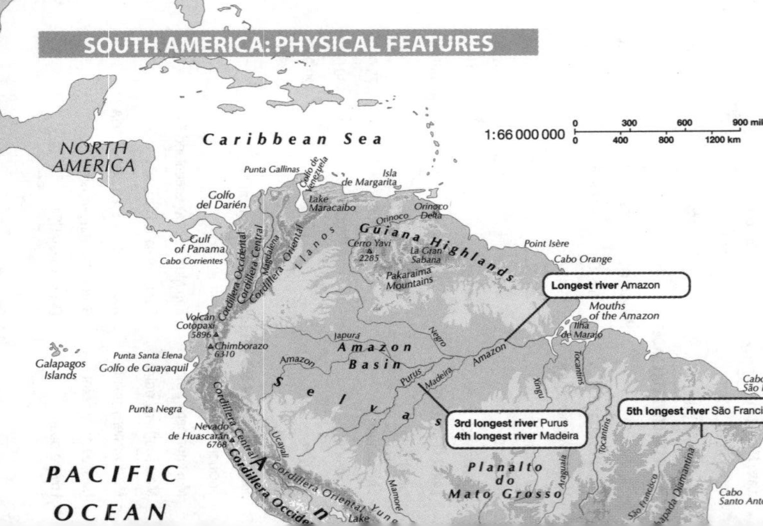
SOUTH AMERICA: PHYSICAL FEATURES
1:66 000 000
0 300 600 900 miles
0 400 800 1200 km
NORTH AMERICA
Caribbean Sea
PACIFIC OCEAN
Galapagos Islands
Golfo del Darién
Gulf of Panama
Cabo Corrientes
Punta Gallinas
Golfo de Venezuela
Lake Maracaibo
Isla de Margarita
Orinoco Delta
Orinoco
Guiana Highlands
Cerro Yavi 2285
La Gran Sabana
Pakaraima Mountains
Llanos
Cordillera Occidental
Cordillera Central
Magdalena
Cordillera Oriental
Volcán Cotopaxi 5896
Chimborazo 6310
Punta Santa Elena
Golfo de Guayaquil
Punta Negra
Nevado de Huascarán 6768
Cordillera Central
Cordillera Occidental
Cordillera Oriental
Lake
Ucayali
Amazon
Japurá
Negro
Amazon Basin
Selvas
Purus
Madeira
Mamoré
Point Isère
Cabo Orange
Mouths of the Amazon
Ilha de Marajó
Tocantins
Xingu
Araguaia
Planalto do Mato Grosso
São Francisco
Cabo de São Roque
Cabo Santo Antonio
Longest river Amazon
3rd longest river Purus
4th longest river Madeira
5th longest river São Francisco

2nd largest desert Atacama
2nd highest mountain Nevado Ojos del Salado
3rd highest mountain Cerro Bonete
4th highest mountain Cerro Pissis
5th highest active volcano Tupungatito
2nd largest island Isla de Chiloé
Largest desert Patagonian
Largest island Tierra del Fuego
Highest mountain Cerro Aconcagua
5th highest mountain Cerro Tupungato
2nd longest river Río del la Plata-Paraná
3rd largest island East Falkland
4th largest island West Falkland
Punta Ballena
Volcán Llullaillaco
Islas de los Desventurados
Nevado Ojos del Salado 6908
Cerro Pissis 6858
Cerro Bonete 6872
Salado
Paraná
Uruguay
Serra do Mar
Ilha de São Sebastião
Cerro Aconcagua 6959
6800 Cerro Tupungato
Tupungatito 6000
San Felix
San Ambrosio
Archipiélago Juan Fernández
Pampas
Río de la Plata
Punta Norte
Punta Sur
Punta Lavapié
Colorado
Negro
ATLANTIC
OCEAN
Punta Calera
Bahía Blanca
Golfo San Matías
Isla de Chiloé
Península Valdés
Patagonia
Archipiélago de los Chonos
Golfo de San Jorge
Cabo Tres Puntas
Golfo de Penas
Falkland Islands
Bahía Grande
West Falkland
East Falkland
Strait of Magellan
Isla de los Estados
Isla Grande de Tierra del Fuego
Cape Horn
Drake Passage
South Georgia
Scotia Sea
South Sandwich Islands

See pages 66–67 for more detail.

SOUTH AMERICA: PHYSICAL EXTREMES

HIGHEST MOUNTAINS	Height (m)	Height (ft)	Location
Cerro Aconcagua	6 960	22 834	Argentina
Nevado Ojos del Salado	6 908	22 664	Argentina/Chile
Cerro Bonete	6 872	22 546	Argentina
Cerro Pissis	6 858	22 500	Argentina
Cerro Tupungato	6 800	22 309	Argentina/Chile

LARGEST ISLANDS	Area (sq km)	Area (sq miles)
Tierra del Fuego	47 000	18 147
Isla de Chiloé	8 394	3 240
East Falkland	6 760	2 610
West Falkland	5 413	2 090

LONGEST RIVERS	Length (km)	Length (miles)
Amazon	6 516	4 049
Río de la Plata-Parana	4 500	2 796
Purus	3 218	2 000
Madeira	3 200	1 988
São Francisco	2 900	1 802

LARGEST LAKE	Area (sq km)	Area (sq miles)
Lake Titicaca	8 340	3 220

HIGHEST ACTIVE VOLCANOES

	Height (m)	Height (ft)	Location
Volcán Llullaillaco	6 723	22 057	Chile
San Pedro	6 199	20 338	Chile
Volcán Aracar	6 082	19 954	Argentina
Guallatiri	6 060	19 882	Chile
Tupungatito	6 000	19 685	Chile

LARGEST DESERTS	Area (sq km)	Area (sq miles)
Patagonian	673 000	260 000
Atacama	140 000	54 000

LAND AREA

Total land area 17 815 420 sq km / 6 878 572 sq miles

Most northerly point	Punta Gallinas, Colombia
Most southerly point	Cape Horn, Chile
Most westerly point	Galagapos Islands, Ecuador
Most easterly point	Ilhas Martin Vas, Atlantic Ocean
Lowest point	Península Valdés, Argentina

THE OCEANS AND POLAR REGIONS

THE OCEANS

COMPARATIVE SIZES OF THE OCEANS

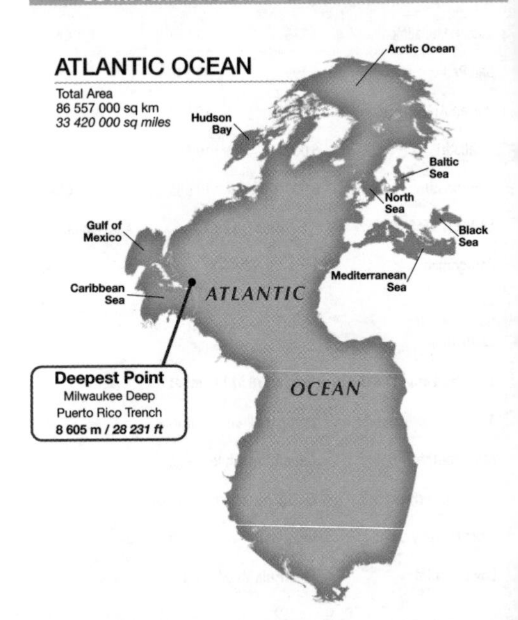

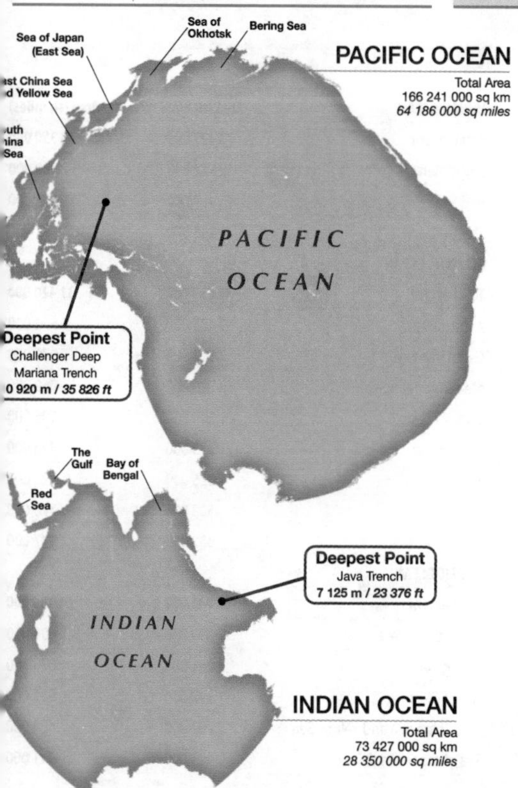
Sea of Okhotsk
Bering Sea
Sea of Japan (East Sea)
PACIFIC OCEAN
Total Area
166 241 000 sq km
64 186 000 sq miles
ast China Sea
d Yellow Sea
uth
hina
Sea
PACIFIC OCEAN
Deepest Point
Challenger Deep
Mariana Trench
0 920 m / 35 826 ft
The Gulf
Bay of Bengal
Red Sea
Deepest Point
Java Trench
7 125 m / 23 376 ft
INDIAN OCEAN
INDIAN OCEAN
Total Area
73 427 000 sq km
28 350 000 sq miles

DIMENSIONS OF OCEANS AND MAJOR SEAS

INDIAN OCEAN	Area (sq km)	Area (sq miles)
Total extent	**73 427 000**	**28 350 000**
Bay of Bengal	2 172 000	839 000
Red Sea	453 000	175 000
The Gulf	238 000	92 000
ATLANTIC OCEAN		
Total extent	**86 557 000**	**33 420 000**
Arctic Ocean	9 485 000	3 662 000
Caribbean Sea	2 512 000	970 000
Mediterranean Sea	2 510 000	969 000
Gulf of Mexico	1 544 000	596 000
Hudson Bay	1 233 000	476 000
North Sea	575 000	222 000
Black Sea	508 000	196 000
Baltic Sea	382 000	147 000
PACIFIC OCEAN		
Total extent	**166 241 000**	**64 186 000**
South China Sea	2 590 000	1 000 000
Bering Sea	2 261 000	873 000
Sea of Okhotsk	1 392 000	537 000
East China Sea and Yellow Sea	1 202 000	464 000
Sea of Japan (East Sea)	1 013 000	391 000

Deepest point (m)	Deepest point (ft)
Java Trench 7 125	**23 376**
4 500	14 763
3 040	9 973
73	239
Milwaukee Deep 8 605	**28 231**
5 450	17 880
7 680	25 196
5 121	16 800
3 504	11 495
259	849
661	2 168
2 245	7 365
460	1 509
Challenger Deep 10 920	**35 826**
5 514	18 090
4 150	13 615
3 363	11 033
2 717	8 913
3 743	12 280

ATLANTIC AND INDIAN OCEANS

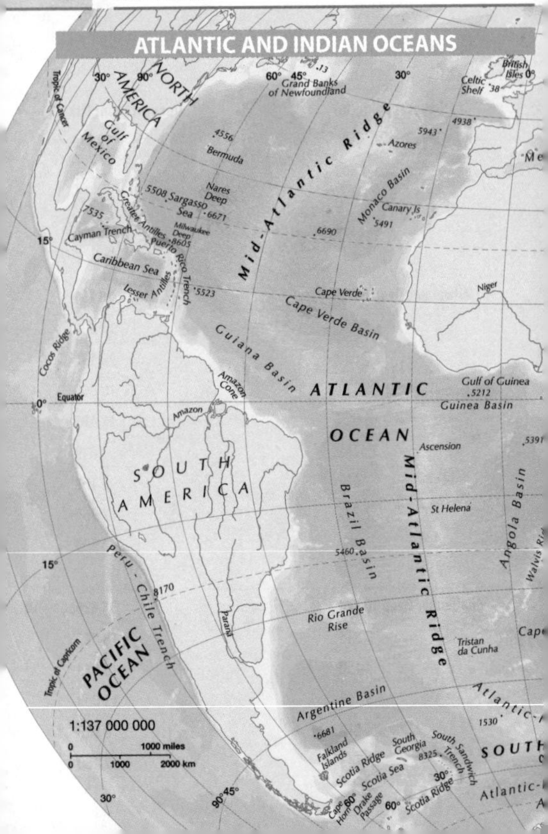

NORTH AMERICA
Gulf of Mexico
Tropic of Cancer
Grand Banks of Newfoundland
Celtic Shelf
British Isles
Bermuda
Azores
Mid-Atlantic Ridge
Nares Deep
Sargasso Sea
Monaco Basin
Canary Is
Greater Antilles
Milwaukee Deep
Puerto Rico Trench
Cayman Trench
Caribbean Sea
Lesser Antilles
Cape Verde
Cape Verde Basin
Niger
Cocos Ridge
Guiana Basin
Amazon Cone
Equator
Amazon
ATLANTIC OCEAN
Gulf of Guinea
Guinea Basin
Ascension
SOUTH AMERICA
Brazil Basin
St Helena
Angola Basin
Walvis Ridge
Peru - Chile Trench
Paraná
Rio Grande Rise
Tristan da Cunha
Tropic of Capricorn
PACIFIC OCEAN
Argentine Basin
Falkland Islands
Scotia Ridge
South Georgia
South Sandwich Trench
Scotia Sea
Drake Passage
Cape Horn
1:137 000 000
0 1000 miles
0 1000 2000 km

ASIA
Irtysh
Aral Sea
Caspian Sea
Black Sea
Sea
Yellow River
Yellow Sea
East China Sea
7460
9156
9695
Yangtze
Indus
Ganges
The Gulf
Red Sea
Nile
Ganges Cone
Bay of Bengal
Andaman Islands
Andaman Basin
4267
South China Sea
Arabian Sea
Gulf of Aden
CA
Carlsberg Ridge
Chagos-Laccadive Ridge
Chagos Trench
5060
Somali Basin
Sumatra
Borneo
Mascarene Ridge
Vema Trench
6402
Mid-Indian Basin
Ninetyeast Ridge
Laut Jawa
Java
Java Trench (Sunda Trench)
7125
Comoros
Mozambique Channel
Zambezi
Madagascar
Mascarene Basin
Mid-Indian Ridge
INDIAN OCEAN
West Australian Basin
North Australian Basin
Madagascar Basin
6400
1924
Southwest Indian Ridge
Natal Basin
1207
Mozambique Ridge
549
2067
Broken Plateau
Perth Basin
Southeast Indian Ridge
AUSTRALIA
Diamantina Deep
7102
6602
6195
Îles Kerguélen
230
McDonald Islands
Heard Island
6972
Kerguelen Plateau
Australian-Antarctic Basin
Great Australian Bight
South Australian Basin
5670
Davis Sea
186
30°
60°
90°
45°
120°
15°
0°

PACIFIC OCEAN

120°
90°
60°
30°
Gulf
Alaska
Tufts Abyssal Plain
NORTH AMERICA
Mississippi
4556
Bermuda
ATLANTIC
OCEAN
Tropic of Cancer
Gulf
of
Mexico
Greater Antilles
Milwaukee
Deep
8605
Puerto Rico Trench
15°
Cayman Trench
Caribbean Sea
Lesser
Antilles
Middle America Trench
6662
Guatemala
Basin
7022
EAN
Basin
Cocos Ridge
East Pacific Rise
Equator 0°
Galapagos Is
Gallego
Rise
Galapagos
Rise
Amazon
Marquesas
Islands
Tuamotu Archipelago
1929
SOUTH
AMERICA
Islands
Peru-Chile Trench
1344
A
8170
15°
Roggevee
Basin
Tropic of Capricorn
Paraná
Pacific - Antarctic Ridge
Chile Rise
Mornington
Abyssal
Plain
30°
Southeast Pacific Basin
Antarctic Circle
120°
90°
60°
60°
45°

ARCTIC OCEAN

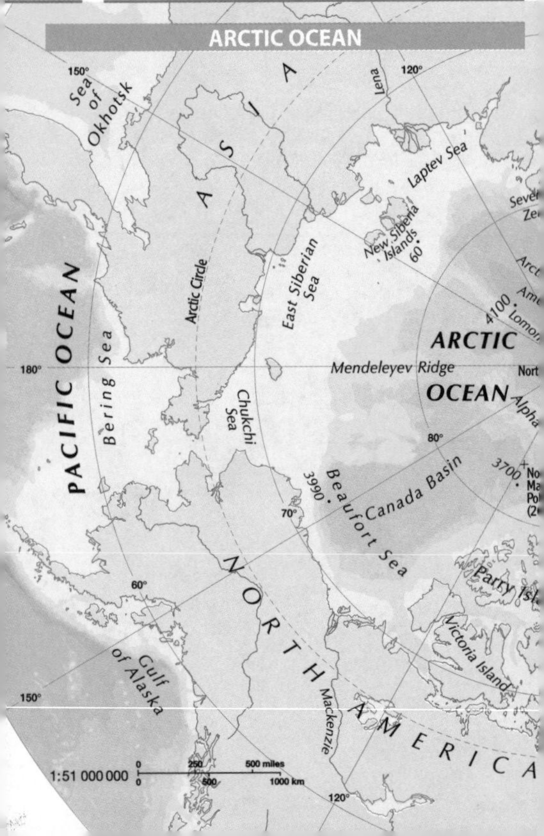

ASIA
Sea of Okhotsk
Lena
Laptev Sea
New Siberia Islands
60
East Siberian Sea
Arctic Circle
PACIFIC OCEAN
Bering Sea
Chukchi Sea
ARCTIC OCEAN
Mendeleyev Ridge
4100
3700
3990
Canada Basin
Beaufort Sea
Parry Isl
Victoria Island
NORTH AMERICA
Mackenzie
Gulf of Alaska
150°
120°
180°
80°
70°
60°
150°
120°
1:51 000 000
0 250 500 miles
0 500 1000 km

Kara Sea
Zemlya Novaya
Barents Sea
Frantsa-Iosifa Zemlya
Spitsbergen
3910
Ridge
EUROPE
Baltic Sea
Norwegian Sea
North Sea
Greenland Sea
3884
Greenland Basin
Jan Mayen
Norwegian Basin
Arctic Circle
3970
Faroe Islands
British Isles
Iceland
Denmark Strait
304
North Geomagnetic Pole (2005)
Greenland
Baffin Bay
Baffin Island
Davis Strait
ATLANTIC OCEAN
60°
30°
0°

POLAR REGIONS

ANTARCTICA

50°
0°
30°
Antarctic Circle
American-Antarctic Ridge
Atlantic-Indian-
Antarctic Basin
60°
100
Thorshavnh
Fimbull
Ice Shelf
Queen Maud
Land
South Sandwich Trench
SOUTHERN OCEAN
30°
70°
Ea
Cape Norvegia
Riiser-Larsen
Ice Shelf
Coats Land
South
Georgia
Scotia Sea
South Orkney
Islands
Scotia Ridge
Weddell Sea
80°
Shackleton
Range
Filchner
Ice Shelf
Berkner
Island
Pensacola Mts
Ronne
Ice Shelf
Vinson
Massif
4897
Ellsworth
Mountains
South Shetland Is
Larsen
Ice Shelf
Palmer Land
Graham Land
Antarctic Peninsula
Alexander
Island
Drake Passage
60°
Marie
Bellingshausen
Sea
Abbot Ice Shelf
Cape
Horn
Peter I
Island
Carney
Island
SOUTH
AMERICA
Amundsen
Amundsen R
Southeast Pacific Basin
90°
120°

60°
90°
120°
150°
150°
180°
Cape Ann
1:51 000 000
0 250 500 miles
0 500 1000 km
Mackenzie Bay
Prince Charles Mountains
Amery Ice Shelf
Davis Sea
Mill Island
SOUTHERN OCEAN
Indian-Antarctic Basin
Vincennes Bay
ntarctica
4000
Wilkes Land
South Geomagnetic Pole (2005)
Cape Morse
Dumont d'Urville Sea
tarctic Mountains
n Maud Mts
Adélie Land
South Magnetic Pole (2005)
George V Land
Ross Ice Shelf
Mount Erebus 3794
Victoria Land
Roosevelt Island
Ross Sea
Balleny Islands
Indian-Antarctic Ridge
Antarctic Circle

POLAR STATISTICS

ANTARCTICA		
Area	**sq km**	**sq miles**
Total land area (excluding ice shelves)	12 093 000	4 669 107
Ice shelves	1 559 000	601 930
Exposed rock	49 000	18 919
Heights	**m**	**ft**
Lowest bedrock elevation (Bentley Subglacial Trench)	- 2 496	- 8 189
Maximum ice thickness (Astrolabe Subglacial Basin)	4 776	15 669
Mean ice thickness (including ice shelves)	1 859	6 099
Volume	**cubic km**	**cubic miles**
Ice sheet including ice shelves	25 400 000	6 094 628
Climate	**°C**	**°F**
Lowest screen temperature (Vostok Station, 21st July 1983)	-89.2	-128.6
Coldest place – Annual mean (Plateau Station)	-56.6	-69.9

THE ARCTIC		
Temperature	**°C**	**°F**
Summer temperature at North Pole	near 0	32
Winter temperature at North Pole	-30	-22
Lowest temperature recorded in the Arctic (Verkhoyansk, northeastern Siberia, 1933)	-68	-90
Sea ice extent	**sq km**	**sq miles**
Minimum sea ice extent in summer	5 000 000	1 930 500
Maximum sea ice extent in winter	16 000 000	6 177 600
Ice thickness	**m**	**ft**
Average sea ice thickness	2	6.6
Maximum ice thickness (Greenland ice sheet)	3 400	11 155
Precipitation	**mm**	**inches**
Average precipitation (mainly snow) in the Arctic basin – rain equivalent	130	5.1
Average precipitation (mainly snow) in the Arctic coastal areas – rain equivalent	260	10.2

THE POLITICAL WORLD

COUNTRIES AND CAPITALS

AL.	ALBANIA	JOR.	JORDAN
A.	ANDORRA	K.	KUWAIT
ARM.	ARMENIA	KYR.	KYRGYZSTAN
AUS.	AUSTRIA	LEB.	LEBANON
AZ.	AZERBAIJAN	LITH.	LITHUANIA
B.	BURUNDI	LUX.	LUXEMBOURG
BE.	BENIN	M.	MACEDONIA
BEL.	BELGIUM	MO.	MOLDOVA
B.H.	BOSNIA-HERZEGOVINA	NETH.	NETHERLANDS
BN.	BAHRAIN	NI.	NIGERIA
BUR.	BURKINA	POL.	POLAND
CAM.	CAMEROON	Q.	QATAR
C.A.R.	CENTRAL AFRICAN REPUBLIC	R.	RWANDA
C.D'I.	CÔTE D'IVOIRE	SLA.	SLOVAKIA
CR.	CROATIA	SLO.	SLOVENIA
CYP.	CYPRUS	S.M.	SERBIA AND
CZ.R.	CZECH REPUBLIC		MONTENEGRO
DEN.	DENMARK	SUR.	SURINAME
EQ.G.	EQUATORIAL GUINEA	SW.	SWITZERLAND
FR.G.	FRENCH GUIANA	T.	TOGO
GEOR.	GEORGIA	TAJIK.	TAJIKISTAN
GER.	GERMANY	TURKM.	TURKMENISTAN
GH.	GHANA	U.A.E.	UNITED ARAB
GUY.	GUYANA		EMIRATES
HUN.	HUNGARY	UZBEK.	UZBEKISTAN
ISR.	ISRAEL

COUNTRIES OF THE WORLD

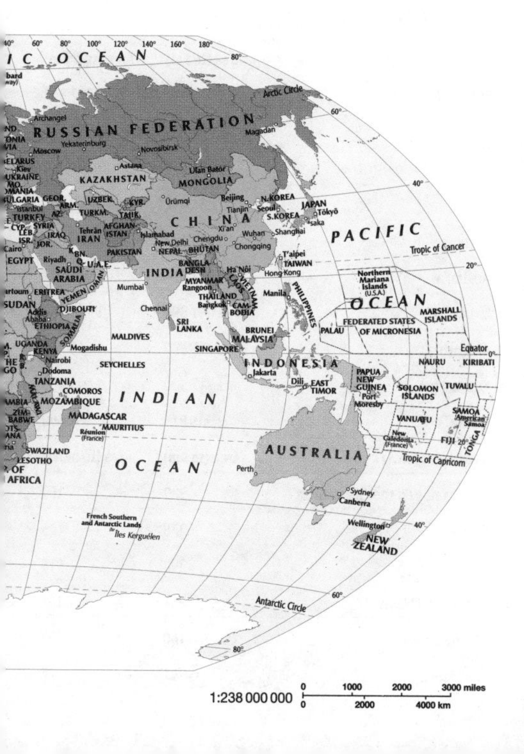

WORLD: LARGEST AND SMALLEST COUNTRIES

COUNTRY SIZE AND CAPITAL EXTREMES

Largest countries	Area (sq km)	Area (sq miles)
Russian Federation	17 075 400	6 592 849
Canada	9 984 670	3 855 103
United States of America	9 826 635	3 794 085
China	9 584 492	3 700 593
Brazil	8 514 879	3 287 613
Australia	7 692 024	2 969 907
India	3 064 898	1 183 364
Argentina	2 766 889	1 068 302
Kazakhstan	2 717 300	1 049 155
Sudan	2 505 813	967 500
Smallest countries	**Area (sq km)**	**Area (sq miles)**
St Vincent and the Grenadines	389	150
Grenada	378	146
Maldives	298	115
St Kitts and Nevis	261	101
Liechtenstein	160	62
San Marino	61	24

Smallest countries	Area (sq km)	Area (sq miles)
Tuvalu	25	10
Nauru	21	8
Monaco	2	1
Vatican City	0.5	0.2
World's national capital extremes		
Most populated	Tōkyō, Japan	35 327 000
Least populated	Yaren, Nauru	600
Highest	La Paz, Bolivia	3 636m / 11 910 ft
Lowest	Manama, Bahrain and and Male, Maldives	0.9 m / 3 ft
Furthest north	Reykjavík, Iceland	64° 08'N
Furthest south	Wellington, New Zealand	41° 18'S
Furthest east	Funafuti, Tuvalu	179° 13'E
Furthest west	Nuku'alofa, Tonga	175° 12'W
Joint national capitals		**Country**
Amsterdam/The Hague		Netherlands
Kuala Lumpur/Putrajaya		Malaysia
La Paz/Sucre		Bolivia
Pretoria/Cape Town		South Africa

WORLD: LARGEST AND SMALLEST COUNTRIES

LARGEST COUNTRIES	Population
China	1 323 345 000
India	1 103 371 000
United States of America	298 213 000
Indonesia	222 781 000
Brazil	186 405 000
Pakistan	157 935 000
Russian Federation	143 202 000
Bangladesh	141 822 000
Nigeria	131 530 000
Japan	128 085 000
SMALLEST COUNTRIES	
Andorra	67 000
Marshall Islands	62 000
St Kitts and Nevis	43 000
Liechtenstein	35 000
Monaco	35 000
San Marino	28 000
Palau	20 000
Nauru	14 000
Tuvalu	10 000
Vatican City	552

HIGHEST POPULATION DENSITIES

	People per sq km	People per sq mile
Monaco	17 500.0	35 000.0
Singapore	6 770.0	17 514.2
Malta	1 272.2	3 295.1
Maldives	1 104.0	2 860.9
Vatican City	1 104.0	2 860.9
Bahrain	1 052.1	2 722.8
Bangladesh	984.9	2 550.8
Nauru	666.7	1 750.0
Taiwan	631.8	1 636.3
Barbados	627.9	1 626.5

LOWEST POPULATION DENSITIES

	People per sq km	People per sq mile
Guyana	3.5	9.0
Libya	3.3	8.6
Canada	3.2	8.4
Botswana	3.0	7.9
Mauritania	3.0	7.9
Iceland	2.9	7.4
Suriname	2.7	7.1
Australia	2.6	6.8
Namibia	2.5	6.4
Mongolia	1.7	4.4

EUROPE: COUNTRIES

NORTH AMERICA

Jan Mayen (Norway)

Reykjavík ICELAND

Norwegian Sea

Tórshavn Faroe Islands (Denmark)

ATLANTIC OCEAN

Bergen Oslo NORWAY

Glasgow Edinburgh North Sea

Belfast UNITED KINGDOM

IRELAND Dublin

Manchester Birmingham Cardiff London

Aalborg DENMARK Copenhagen

Hamburg Berlin

NETH. The Hague Amsterdam Rhine Essen

Brussels BELGIUM GERMANY

English Channel Channel Islands (U.K.)

LUX. Luxembourg Frankfurt am Main

Seine Paris Danube

Nantes Loire Orléans Strasbourg Munich

Zürich LIE. Vaduz

Bay of Biscay FRANCE Bern SW. Geneva

Lyon Milan Ljubljana

Bordeaux Turin Po SAN MARINO

Oporto Andorra la Vella ANDORRA Marseille MONACO

PORTUGAL Lisbon Tagus Madrid SPAIN

Barcelona Corsica Vatican City Rome

Palma de Mallorca Balearic Islands Sardinia Naples Tyrrhenian Sea

Valencia Seville Cartagena

Cádiz Gibraltar (U.K.)

Mediterranean Palermo Valletta MALTA

Azores (Portugal)

Madeira (Portugal)

AFRICA

AL.	ALBANIA
B.H.	BOSNIA-HERZEGOVINA
CR.	CROATIA
CZ.R.	CZECH REPUBLIC
HUN.	HUNGARY
LIE.	LIECHTENSTEIN
LUX.	LUXEMBOURG
M.	MACEDONIA
NETH.	NETHERLANDS
S.M.	SERBIA AND MONTENEGRO
SW.	SWITZERLAND

Novaya Zemlya
Barents Sea
Ostrov Kolguyev
Vorkuta
Pechora
Kola Peninsula
White Sea
Archangel
RUSSIAN FEDERATION
Severnaya Dvina
Lake Ladoga
FINLAND
Perm'
Izhevsk
Helsinki
Gulf of Finland
St Petersburg
Tallinn
ESTONIA
Yaroslavl'
Volga
Nizhniy Novgorod
Kazan'
Ufa
LATVIA
Riga
Moscow
Ul'yanovsk
Samara
Orenburg
ASIA
Tula
Vilnius
Minsk
BELARUS
Saratov
Voronezh
Homyel'
Brest
Dnieper
Volgograd
Kiev
Rivne
Kharkiv
UKRAINE
Donets'k
Don
Astrakhan'
L'viv
Dniester
Dnipropetrovs'k
Rostov-na-Donu
MOLDOVA
Krasnodar
Grozny
Caspian Sea
Odesa
Caucasus
ROMANIA
Bucharest
Constan&a
Danube
Black Sea
BULGARIA
Sofia
Skopje
Istanbul
TURKEY
Thessaloniki
Aegean Sea
GREECE
Athens
Crete
Sea
1:46 500 000
0 150 300 450 miles
0 300 600 km

EUROPE: LARGEST AND SMALLEST COUNTRIES

LARGEST COUNTRIES	Area (sq km)	Area (sq miles)
Russian Federation	17 075 400	6 592 849
Ukraine	603 700	233 090
France	543 965	210 026
Spain	504 782	194 897
Sweden	449 964	173 732
Germany	357 022	137 847
Finland	338 145	130 559
Norway	323 878	125 050
Poland	312 683	120 728
Italy	301 245	116 311
SMALLEST COUNTRIES		
Albania	28 748	11 100
Macedonia (F.Y.R.O.M.)	25 713	9 928
Slovenia	20 251	7 819
Luxembourg	2 586	998
Andorra	465	180
Malta	316	122
Liechtenstein	160	62
San Marino	61	24
Monaco	2	1
Vatican City	0.5	0.2
LARGEST COUNTRIES		**Population**
Russian Federation		143 202 000
Germany		82 689 000
France		60 496 000

United Kingdom		59 668 000
Italy		58 093 000
Ukraine		46 481 000
Spain		43 064 000
Poland		38 530 000
Romania		21 711 000
Netherlands		16 299 000
SMALLEST COUNTRIES		**Population**
Slovenia		1 967 000
Estonia		1 330 000
Luxembourg		465 000
Malta		402 000
Iceland		295 000
Andorra		67 000
Liechtenstein		35 000
Monaco		35 000
San Marino		28 000
Vatican City		552
EUROPE'S CAPITALS		
Largest capital (population)	Paris, France	9 854 000
Smallest capital (population)	Vatican City	472
Most northerly capital	Reykjavík, Iceland	64° 39'N
Most southerly capital	Valletta, Malta	35° 54'N
Highest capital (above sea level)	Andorra la Vella, Andorra	1 029 metres 3 376 feet

ASIA: COUNTRIES

Barents Sea
EUROPE
Mediterranean Sea
Moscow
Nizhniy Novgorod
Volga
Samara
Ural Mountains
RUSSI
Yekaterinburg
Ural'sk
Omsk
Black Sea
Ankara
TURKEY
Adana
GEORGIA
T'bilisi
ARMENIA
Yerevan
AZERBAIJAN
Baku
Caspian Sea
Aral Sea
Astana
KAZAKHSTAN
Lake Balkhash
Nicosia
CYPRUS
LEBANON
Beirut
SYRIA
Damascus
Jerusalem
ISRAEL
Ammān
JORDAN
Tabrīz
UZBEKISTAN
TURKMENISTAN
Bishkek
Almaty
Tashkent
KYRGYZSTAN
Dushanbe
TAJIKISTAN
AFRICA
Baghdād
IRAQ
Tehrān
Ashgabat
IRAN
Herāt
Kābul
AFGHANISTAN
Islamabad
KUWAIT
Kuwait
The Gulf
Shīrāz
Kandahar
Lahore
Red Sea
Jeddah
BAHRAIN
Riyadh
Manama
Mecca
QATAR
Doha
SAUDI ARABIA
Dubai
Abu Dhabi
U.A.E.
Muscat
PAKISTAN
Delhi
New Delhi
Hyderabad
Karachi
Mount Everest
NEPAL
Agra
Kathmandu
Allahabad
Ganges
Şan'ā'
YEMEN
OMAN
Ahmadabad
Aden
Arabian Sea
INDIA
Mumbai
Socotra
Hyderabad
Bangalore
Chennai
Laccadive Islands
Madurai
Sri Jayewardenepura Kotte
Colombo
SRI LANKA
MALDIVES
Male
INDIAN OCEAN
British Indian Ocean Territory
1:103 000 000
0 500 1000 1500 miles
0 1000 2000 km

TIC OCEAN
Bering Sea
Magadan
Lena
Sea of Okhotsk
Petropavlovak-Kamchatskiy
DERATION
Lake Baikal
rkutsk
Ulan Bator
ONGOLIA
Harbin
Vladivostock
Sea of Japan (East Sea)
Sapporo
Hakodate
JAPAN
Tōkyō
Shenyang
NORTH KOREA
P'yŏngyang
Beijing
Dalian
Yellow River
Tianjin
Seoul
SOUTH KOREA
Ōsaka
Hiroshima
Fukuoka
Yellow Sea
PACIFIC OCEAN
Lanzhou
INA
Xi'an
Nanjing
Shanghai
Hangzhou
Yangtze
East China Sea
Chengdu
Wuhan
Chongqing
T'aipei
TAIWAN
Kaoshiung
Kunming
Liuzhou
Nanning
Guangzhou
Hong Kong
Luzon Strait
Ha Nôi
Hai Phong
LAOS
VIETNAM
Vientiane
Quezon City
PHILIPPINES
Manila
South China Sea
THAILAND
Bangkok
CAMBODIA
PALAU
Koror
Phnom Penh
Hô Chi Minh
Davao
Bandar Seri Begawan
Kota Kinabalu
Celebes Sea
MALAYSIA
BRUNEI
Jayapura
Kuala Lumpur
Putrajaya
Kuching
SINGAPORE
Singapore
Borneo
Pontianak
New Guinea
Sumatra
INDONESIA
Palembang
Banjarmasin
Laut Banda
OCEANIA
Jakarta
Laut Jawa
Makassar
EAST TIMOR
Dili
Bandung
Surabaya
Semarang
Java

ASIA: LARGEST AND SMALLEST COUNTRIES

LARGEST COUNTRIES	Area (sq km)	Area (sq miles)
Russian Federation	17 075 400	6 592 849
China	9 584 492	3 700 593
India	3 064 898	1 183 364
Kazakhstan	2 717 300	1 049 155
Saudi Arabia	2 200 000	849 425
Indonesia	1 919 445	741 102
Iran	1 648 000	636 296
Mongolia	1 565 000	604 250
Pakistan	803 940	310 403
Turkey	779 452	300 948
SMALLEST COUNTRIES		
Kuwait	17 818	6 880
East Timor	14 874	5 743
Qatar	11 437	4 416
Lebanon	10 452	4 036
Cyprus	9 251	3 572
Brunei	5 765	2 226
Bahrain	691	267
Singapore	639	247
Palau	497	192
Maldives	298	115
LARGEST COUNTRIES		**Population**
China		1 323 345 000
India		1 103 371 000

LARGEST COUNTRIES		Population
Indonesia		222 781 000
Pakistan		157 935 000
Russian Federation		143 202 000
Bangladesh		141 822 000
Japan		128 085 000
Vietnam		84 238 000
Philippines		83 054 000
Turkey		73 193 000
SMALLEST COUNTRIES		
Mongolia		2 646 000
Oman		2 567 000
Bhutan		2 163 000
East Timor		947 000
Cyprus		835 000
Qatar		813 000
Bahrain		727 000
Brunei		374 000
Maldives		329 000
Palau		20 000
ASIA'S CAPITALS		
Largest capital (population)	Tōkyō, Japan	35 327 000
Smallest capital (population)	Koror, Palau	14 000
Most northerly capital	Astana, Kazakhstan	51° 10'N
Most southerly capital	Dili, East Timor	8° 35'S
Highest capital	Thimphu, Bhutan	2 423 metres 7 949 feet

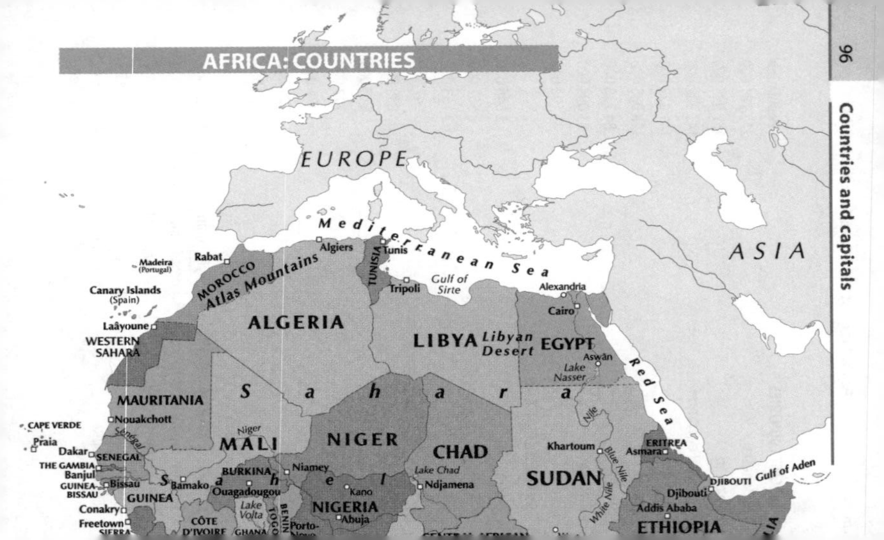
AFRICA: COUNTRIES
EUROPE
ASIA
Mediterranean Sea
Red Sea
Gulf of Aden
Gulf of Sirte
Atlas Mountains
Sahara
Sahel
Libyan Desert
Lake Nasser
Lake Chad
Lake Volta
Nile
Blue Nile
White Nile
Niger
Senegal
Madeira (Portugal)
Canary Islands (Spain)
CAPE VERDE
Praia
WESTERN SAHARA
Laâyoune
MOROCCO
Rabat
ALGERIA
Algiers
TUNISIA
Tunis
LIBYA
Tripoli
EGYPT
Cairo
Alexandria
Aswân
SUDAN
Khartoum
ERITREA
Asmara
DJIBOUTI
Djibouti
ETHIOPIA
Addis Ababa
CHAD
Ndjamena
NIGER
Niamey
NIGERIA
Kano
Abuja
BENIN
Porto-
TOGO
GHANA
BURKINA
Ouagadougou
MALI
Bamako
MAURITANIA
Nouakchott
SENEGAL
Dakar
THE GAMBIA
Banjul
GUINEA-BISSAU
Bissau
GUINEA
Conakry
Freetown
CÔTE D'IVOIRE

Sao Tome
GABON
REPUBLIC
OF THE
CONGO
RWANDA
Lake
Victoria
Nairobi
5199
Kigali
Kilimanjaro
5892
Brazzaville
Kinshasa
Bujumbura
BURUNDI
Victoria
ATLANTIC
OCEAN
Dodoma
Zanzibar Island
Lake
Tanganyika
TANZANIA
Dar es Salaam
SEYCHELLES
Luanda
Ascension
Aldabra
Islands
Lake
Nyasa
ANGOLA
Lubumbashi
Moroni
COMOROS
St Helena and
Dependencies
(U.K.)
Huambo
Lilongwe
MALAWI
Dzaoudzi
Mayotte
(France)
ZAMBIA
Lusaka
Zambezi
Nampula
St Helena
Cubango
MOZAMBIQUE
Harare
ZIMBABWE
MADAGASCAR
Mozambique
Channel
Antananarivo
MAURITIUS
Port Louis
St-Denis
Réunion
(France)
Namib Desert
Okavango
Delta
Bulawayo
BOTSWANA
NAMIBIA
Windhoek
Limpopo
Gaborone
Pretoria
Maputo
Johannesburg
Mbabane
SWAZILAND
Orange
Maseru
LESOTHO
Durban
INDIAN
OCEAN
REPUBLIC OF
SOUTH AFRICA
Cape Town
Port Elizabeth
Cape of
Good Hope
Cape Agulhas
1:79 500 000
0 500 1000 miles
0 500 1000 1500 km

AFRICA: LARGEST AND SMALLEST COUNTRIES

LARGEST COUNTRIES	Area (sq km)	Area (sq miles)
Sudan	2 505 813	967 500
Algeria	2 381 741	919 595
Democratic Republic of the Congo	2 345 410	905 568
Libya	1 759 540	679 362
Chad	1 284 000	495 755
Niger	1 267 000	489 191
Angola	1 246 700	481 354
Mali	1 240 140	478 821
Republic of South Africa	1 219 090	470 693
Ethiopia	1 133 880	437 794
SMALLEST COUNTRIES		
Burundi	27 835	10 747
Rwanda	26 338	10 169
Djibouti	23 200	8 958
Swaziland	17 364	6 704
The Gambia	11 295	4 361
Cape Verde	4 033	1 557
Mauritius	2 040	788
Comoros	1 862	719
São Tomé and Príncipe	964	372
Seychelles	455	176
LARGEST COUNTRIES		**Population**
Nigeria		131 530 000
Ethiopia		77 431 000
Egypt		74 033 000

LARGEST COUNTRIES		Population
Democratic Republic of the Congo		57 549 000
Republic of South Africa		47 432 000
Tanzania		38 329 000
Sudan		36 233 000
Kenya		34 256 000
Algeria		32 854 000
Morocco		31 478 000
SMALLEST COUNTRIES		
The Gambia		1 517 000
Gabon		1 384 000
Mauritius		1 245 000
Swaziland		1 032 000
Comoros		798 000
Djibouti		793 000
Cape Verde		507 000
Equatorial Guinea		504 000
São Tomé and Príncipe		157 000
Seychelles		81 000
AFRICA'S CAPITALS		
Largest capital (population)	Cairo, Egypt	11 146 000
Smallest capital (population)	Victoria, Seychelles	30 000
Most northerly capital	Tunis, Tunisia	36° 46'N
Most southerly capital	Cape Town, Republic of South Africa	33° 57'S
Highest capital	Addis Ababa, Ethiopia	2 408 metres 7 900 feet

OCEANIA: COUNTRIES

Wake Island (U.S.A.)
Pagan
Northern Mariana Islands (U.S.A.)
Saipan
Capitol Hill
Guam (U.S.A.)
Hagåtña
MARSHALL ISLANDS
Yap
Gaferut
Chuuk
Pohnpei
Palikir
Caroline Islands
Kosrae
FEDERATED STATES OF MICRONESIA
Majuro
Gilbert Islands
Yaren
NAURU
Kingsmill Group
ASIA
Rabaul
New Ireland
Mount Wilhelm 4509
New Guinea
PAPUA NEW GUINEA
New Britain
Bougainville I.
SOLOMON ISLANDS
Solomon Sea
Honiara
Malaita
Santa Cruz Islands
Arafura Sea
Torres Strait
Port Moresby
Banks Islands
VANUATU
Espíritu Santo
Malakula
Éfaté
Port Vila
Timor Sea
Darwin
Gulf of Carpentaria
Coral Sea Islands Territory (Australia)
Coral Sea
Cairns
New Caledonia (France)
Nouméa
Îles Loyauté
Lake Argyle
Townsville
Cape Lévêque
INDIAN OCEAN
Broome
AUSTRALIA
Uluru 867
Alice Springs
Brisbane
North West Cape
Norfolk Island (Australia)
Lake Eyre
Darling
Lord Howe Island (Australia)
North Cape
Lake Torrens
Sydney
Canberra
Kalgoorlie
Great Australian Bight
Adelaide
Murray
Mount Kosciuszko 2229
Melbourne
Tasman Sea
North Island
Wellington
Perth
Kangaroo Island
Bass Strait
Cape Leeuwin
Tasmania
Hobart
Aoraki 3754
South Island
Stewart Island
Auckland Islands (N.Z.)
Campbell Island (N.Z.)
Macquarie Island (Australia)

Hawaiian Islands (U.S.A.)

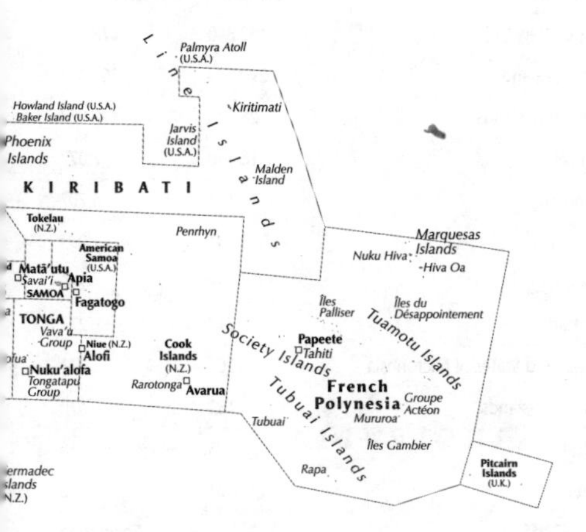
PACIFIC OCEAN
Line Islands
Palmyra Atoll (U.S.A.)
Kiritimati
Howland Island (U.S.A.)
Baker Island (U.S.A.)
Jarvis Island (U.S.A.)
Phoenix Islands
Malden Island
KIRIBATI
Tokelau (N.Z.)
Penrhyn
Marquesas Islands
Nuku Hiva
Hiva Oa
American Samoa (U.S.A.)
Matā'utu
Savai'i
Apia
SAMOA
Fagatogo
TONGA
Vava'u Group
Niue (N.Z.)
Alofi
Nuku'alofa
Tongatapu Group
Cook Islands (N.Z.)
Rarotonga
Avarua
Îles Palliser
Îles du Désappointement
Society Islands
Papeete
Tahiti
Tuamotu Islands
French Polynesia
Tubuai Islands
Groupe Actéon
Mururoa
Tubuai
Îles Gambier
Rapa
Pitcairn Islands (U.K.)
ermadec slands N.Z.)
Chatham Islands (N.Z.)

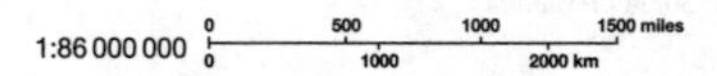
1:86 000 000
0 500 1000 1500 miles
0 1000 2000 km

OCEANIA: LARGEST AND SMALLEST COUNTRIES

COUNTRIES	Area (sq km)	Area (sq miles)
Australia	7 692 024	2 969 907
Papua New Guinea	462 840	178 704
New Zealand	270 534	104 454
Solomon Islands	28 370	10 954
Fiji	18 330	7 077
Vanuatu	12 190	4 707
Samoa	2 831	1 093
Tonga	748	289
Kiribati	717	277
Federated States of Micronesia	701	271
Marshall Islands	181	70
Tuvalu	25	10
Nauru	21	8

COUNTRIES	Population
Australia	20 155 000
Papua New Guinea	5 887 000
New Zealand	4 028 000
Fiji	848 000
Solomon Islands	478 000

COUNTRIES		Population
Vanuatu		211 000
Samoa		185 000
Federated States of Micronesia		110 000
Tonga		102 000
Kiribati		99 000
Marshall Islands		62 000
Nauru		14 000
Tuvalu		10 000
OCEANIA'S CAPITALS		
Largest capital (population)	Canberra, Australia	387 000
Smallest capital (population)	Vaiaku, Tuvalu	5 100
Most northerly capital	Delap-Uliga-Djarrit, Marshall Islands	7° 7'N
Most southerly capital	Wellington, New Zealand	41° 18'S
Highest capital	Canberra, Australia	581 metres 1 906 feet

NORTH AMERICA: COUNTRIES

Queen Charlotte Islands

PACIFIC OCEAN

Fraser
Edmonton
Vancouver Island
Vancouver
Calgary
Seattle
Portland
Great Salt Lake
Salt Lake City
San Francisco
UNITED S
Colorado
Phoenix
El Paso

Kauai
Oahu
Hawaiian Islands (U.S.A.)
Maui
Hawaii

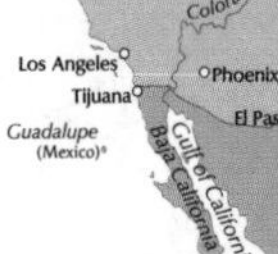

Guadalajara

1:86 000 000
0 500 1000 1500 miles
0 1000 2000 km

Greenland Sea
Denmark Strait
Greenland
Baffin Bay
Davis Strait
Baffin Island
Nuuk
Cape Farewell
Foxe Basin
Hudson Strait
Labrador Sea
Hudson Bay
Belcher Islands
James Bay
Newfoundland
St John's
Île d'Anticosti
Gulf of St Lawrence
St-Pierre
St Pierre and Miquelon (France)
Lake Nipigon
Québec
Halifax
Ottawa
Montréal
Cape Sable
Portland
Great Lakes
Boston
Minneapolis
Toronto
Detroit
Cleveland
New York
Chicago
Columbus
Pittsburgh
Philadelphia
Washington
ATLANTIC OCEAN
Ohio
Cape Hatteras
Bermuda (U.K.)
Mississippi
Atlanta
Jacksonville
Orlando
New Orleans
THE BAHAMAS
Nassau
Miami
Gulf of Mexico
Turks and Caicos Islands (U.K.)
Virgin Islands (U.S.A.)
Virgin Islands (U.K.)
San Juan
Santo Domingo
ST KITTS AND NEVIS
ANTIGUA AND BARBUDA
Guadeloupe (France)
Havana
CUBA
Puerto Rico (U.S.A.)
DOMINICA
Martinique (France)
Mérida
Yucatán
Cayman Islands (U.K.)
Kingston
HAITI
Port-au-Prince
DOMINICAN REPUBLIC
ST LUCIA
BARBADOS
JAMAICA
ST VINCENT AND THE GRENADINES
GRENADA
TRINIDAD AND TOBAGO
BELIZE
Belmopan
Caribbean Sea
Aruba (Neth.)
Netherlands Antilles
HONDURAS
Tegucigalpa
NICARAGUA
Managua
EL SALVADOR
Lake Nicaragua
Panama Canal
San José
COSTA RICA
PANAMA
Panama City
SOUTH AMERICA

NORTH AMERICA: LARGEST AND SMALLEST COUNTRIES

LARGEST COUNTRIES	Area (sq km)	Area (sq miles)
Canada	9 984 670	3 855 103
United States of America	9 826 635	3 794 085
Mexico	1 972 545	761 604
Nicaragua	130 000	50 193
Honduras	112 088	43 277
Cuba	110 860	42 803
Guatemala	108 890	42 043
Panama	77 082	29 762
Costa Rica	51 100	19 730
Dominican Republic	48 442	18 704
SMALLEST COUNTRIES		
The Bahamas	13 939	5 382
Jamaica	10 991	4 244
Trinidad and Tobago	5 130	1 981
Dominica	750	290
St Lucia	616	238
Antigua and Barbuda	442	171
Barbados	430	166
St Vincent and The Grenadines	389	150
Grenada	378	146
St Kitts and Nevis	261	101
LARGEST COUNTRIES		**Population**
United States of America		298 213 000
Mexico		107 029 000
Canada		32 268 000

LARGEST COUNTRIES		Population
Guatemala		12 599 000
Cuba		11 269 000
Dominican Republic		8 895 000
Haiti		8 528 000
Honduras		7 205 000
El Salvador		6 881 000
Nicaragua		5 487 000
SMALLEST COUNTRIES		
Trinidad and Tobago		1 305 000
The Bahamas		323 000
Barbados		270 000
Belize		270 000
St Lucia		161 000
St Vincent and the Grenadines		119 000
Grenada		103 000
Antigua and Barbuda		81 000
Dominica		79 000
St Kitts and Nevis		43 000
NORTH AMERICA'S CAPITALS		
Largest capital (population)	Mexico City, Mexico	19 013 000
Smallest capital (population)	Belmopan, Belize	9 000
Most northerly capital	Ottawa, Canada	45° 25'N
Most southerly capital	Panama City, Panama	8° 56'N
Highest capital	Mexico City, Mexico	2 300 metres 7 546 feet

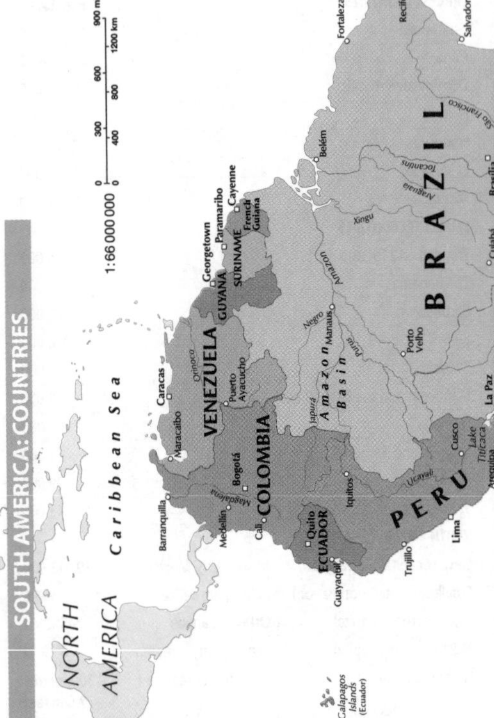
SOUTH AMERICA: COUNTRIES
1:66 000 000
0 300 600 900 miles
0 400 800 1200 km
NORTH AMERICA
Caribbean Sea
Galapagos Islands (Ecuador)
VENEZUELA
GUYANA
SURINAME
French Guiana
COLOMBIA
ECUADOR
PERU
BRAZIL
Barranquilla
Maracaibo
Caracas
Orinoco
Puerto Ayacucho
Georgetown
Paramaribo
Cayenne
Medellín
Bogotá
Magdalena
Cali
Quito
Guayaquil
Iquitos
Japurá
Amazon Basin
Negro
Manaus
Purus
Amazon
Xingu
Araguaia
Tocantins
Belém
Fortaleza
Recife
Salvador
São Francisco
Brasília
Cuiabá
Porto Velho
La Paz
Lake Titicaca
Cusco
Arequipa
Ucayali
Lima
Trujillo

OCEAN
Antofagasta
Islas de los Desventurados
Archipiélago Juan Fernández
San Felix
San Ambrosio
Santiago
Concepción
Puerto Montt
CHILE
ARGENTINA
Salado
Córdoba
Mendoza
Colorado
Neuquén
Negro
Viedma
Comodoro Rivadavia
Punta Arenas
Ushuaia
Isla Grande de Tierra del Fuego
Asunción
Paraná
Uruguay
Concordia
URUGUAY
Buenos Aires
Montevideo
Mar del Plata
Curitiba
São Paulo
Janeiro
Porto Alegre
ATLANTIC
OCEAN
Falkland Islands (U.K)
Stanley
Scotia Sea
South Georgia and South Sandwich Islands (U.K)

SOUTH AMERICA: LARGEST AND SMALLEST COUNTRIES

COUNTRIES	Area (sq km)	Area (sq miles)
Brazil	8 514 879	3 287 613
Argentina	2 766 889	1 068 302
Peru	1 285 216	496 225
Colombia	1 141 748	440 831
Bolivia	1 098 581	424 164
Venezuela	912 050	352 144
Chile	756 945	292 258
Paraguay	406 752	157 048
Ecuador	272 045	105 037
Guyana	214 969	83 000
Uruguay	176 215	68 037
Suriname	163 820	63 251
COUNTRIES		**Population**
Brazil		186 405 000
Colombia		45 600 000
Argentina		38 747 000
Peru		27 968 000

COUNTRIES		Population
Venezuela		26 749 000
Chile		16 295 000
Ecuador		13 228 000
Bolivia		9 182 000
Paraguay		6 158 000
Uruguay		3 463 000
Guyana		751 000
Suriname		449 000
SOUTH AMERICA'S CAPITALS		
Largest capital (population)	Buenos Aires, Argentina	13 349 000
Smallest capital (population)	Sucre, Bolivia	183 000
Most northerly capital	Caracas, Venezuela	10° 28'N
Most southerly capital	Buenos Aires, Argentina	34° 36'S
Highest capital	La Paz, Bolivia	3 630 metres 11 909 feet

STATES AND TERRITORIES

INDEPENDENT COUNTRIES

	Continent	Population
AFGHANISTAN	Asia	29 863 000
ALBANIA	Europe	3 130 000
ALGERIA	Africa	32 854 000
ANDORRA	Europe	67 000
ANGOLA	Africa	15 941 000
ANTIGUA AND BARBUDA	North America	81 000
ARGENTINA	South America	38 747 000
ARMENIA	Asia	3 016 000
AUSTRALIA	Oceania	20 155 000
AUSTRIA	Europe	8 189 000
AZERBAIJAN	Asia	8 411 000
THE BAHAMAS	North America	323 000
BAHRAIN	Asia	727 000
BANGLADESH	Asia	141 822 000
BARBADOS	North America	270 000
BELARUS	Europe	9 755 000
BELGIUM	Europe	10 419 000
BELIZE	North America	270 000
BENIN	Africa	8 439 000

Capital	Area (sq km)	Area (sq miles)
Kābul	652 225	251 825
Tirana	28 748	11 100
Algiers	2 381 741	919 595
Andorra la Vella	465	180
Luanda	1 246 700	481 354
St John's	442	171
Buenos Aires	2 766 889	1 068 302
Erevan	29 800	11 506
Canberra	7 692 024	2 969 907
Vienna	83 855	32 377
Baku	86 600	33 436
Nassau	13 939	5 382
Manama	691	267
Dhaka	143 998	55 598
Bridgetown	430	166
Minsk	207 600	80 155
Brussels	30 520	11 784
Belmopan	22 965	8 867
Porto-Novo	112 620	43 483

	Continent	Population
BHUTAN	Asia	2 163 000
BOLIVIA	South America	9 182 000
BOSNIA-HERZEGOVINA	Europe	3 907 000
BOTSWANA	Africa	1 765 000
BRAZIL	South America	186 405 000
BRUNEI	Asia	374 000
BULGARIA	Europe	7 726 000
BURKINA	Africa	13 228 000
BURUNDI	Africa	7 548 000
CAMBODIA	Asia	14 071 000
CAMEROON	Africa	16 322 000
CANADA	North America	32 268 000
CAPE VERDE	Africa	507 000
CENTRAL AFRICAN REPUBLIC	Africa	4 038 000
CHAD	Africa	9 749 000
CHILE	South America	16 295 000
CHINA	Asia	1 323 345 000
COLOMBIA	South America	45 600 000
COMOROS	Africa	798 000
CONGO	Africa	3 999 000
CONGO, DEMOCRATIC REPUBLIC OF THE	Africa	57 549 000
COSTA RICA	North America	4 327 000

Capital	Area (sq km)	Area (sq miles)
Thimphu	46 620	18 000
La Paz/Sucre	1 098 581	424 164
Sarajevo	51 130	19 741
Gaborone	581 370	224 468
Brasília	8 514 879	3 287 613
Bandar Seri Begawan	5 765	2 226
Sofia	110 994	42 855
Ouagadougou	274 200	105 869
Bujumbura	27 835	10 747
Phnom Penh	181 035	69 884
Yaoundé	475 442	183 569
Ottawa	9 984 670	3 855 103
Praia	4 033	1 557
Bangui	622 436	240 324
Ndjamena	1 284 000	495 755
Santiago	756 945	292 258
Beijing	9 584 492	3 700 593
Bogotá	1 141 748	440 831
Moroni	1 862	719
Brazzaville	342 000	132 047
Kinshasa	2 345 410	905 568
San José	51 100	19 730

	Continent	Population
CÔTE D'IVOIRE	Africa	18 154 000
CROATIA	Europe	4 551 000
CUBA	North America	11 269 000
CYPRUS	Asia	835 000
CZECH REPUBLIC	Europe	10 220 000
DENMARK	Europe	5 431 000
DJIBOUTI	Africa	793 000
DOMINICA	North America	79 000
DOMINICAN REPUBLIC	North America	8 895 000
EAST TIMOR	Asia	947 000
ECUADOR	South America	13 228 000
EGYPT	Africa	74 033 000
EL SALVADOR	North America	6 881 000
EQUATORIAL GUINEA	Africa	504 000
ERITREA	Africa	4 401 000
ESTONIA	Europe	1 330 000
ETHIOPIA	Africa	77 431 000
FIJI	Oceania	848 000
FINLAND	Europe	5 249 000
FRANCE	Europe	60 496 000
GABON	Africa	1 384 000
THE GAMBIA	Africa	1 517 000

Capital	Area (sq km)	Area (sq miles)
Yamoussoukro	322 463	124 504
Zagreb	56 538	21 829
Havana	110 860	42 803
Nicosia	9 251	3 572
Prague	78 864	30 450
Copenhagen	43 075	16 631
Djibouti	23 200	8 958
Roseau	750	290
Santo Domingo	48 442	18 704
Dili	14 874	5 743
Quito	272 045	105 037
Cairo	1 000 250	386 199
San Salvador	21 041	8 124
Malabo	28 051	10 831
Asmara	117 400	45 328
Tallinn	45 200	17 452
Addis Ababa	1 133 880	437 794
Suva	18 330	7 077
Helsinki	338 145	130 559
Paris	543 965	210 026
Libreville	267 667	103 347
Banjul	11 295	4 361

	Continent	Population
GEORGIA	Asia	4 474 000
GERMANY	Europe	82 689 000
GHANA	Africa	22 113 000
GREECE	Europe	11 120 000
GRENADA	North America	103 000
GUATEMALA	North America	12 599 000
GUINEA	Africa	9 402 000
GUINEA-BISSAU	Africa	1 586 000
GUYANA	South America	751 000
HAITI	North America	8 528 000
HONDURAS	North America	7 205 000
HUNGARY	Europe	10 098 000
ICELAND	Europe	295 000
INDIA	Asia	1 103 371 000
INDONESIA	Asia	222 781 000
IRAN	Asia	69 515 000
IRAQ	Asia	28 807 000
IRELAND	Europe	4 148 000
ISRAEL	Asia	6 725 000
ITALY	Europe	58 093 000
JAMAICA	North America	2 651 000

Capital	Area (sq km)	Area (sq miles)
T'bilisi	69 700	26 911
Berlin	357 022	137 849
Accra	238 537	92 100
Athens	131 957	50 949
St George's	378	146
Guatemala City	108 890	42 043
Conakry	245 857	94 926
Bissau	36 125	13 948
Georgetown	214 969	83 000
Port-au-Prince	27 750	10 714
Tegucigalpa	112 088	43 277
Budapest	93 030	35 919
Reykjavík	102 820	39 699
New Delhi	3 064 898	1 183 364
Jakarta	1 919 445	741 102
Tehrān	1 648 000	636 296
Baghdād	438 317	169 235
Dublin	70 282	27 136
Jerusalem*	20 770	8 019
Rome	301 245	116 311
Kingston	10 991	4 244

*De facto capital. Disputed.

	Continent	Population
JAPAN	Asia	128 085 000
JORDAN	Asia	5 703 000
KAZAKHSTAN	Asia	14 825 000
KENYA	Africa	34 256 000
KIRIBATI	Oceania	99 000
KUWAIT	Asia	2 687 000
KYRGYZSTAN	Asia	5 264 000
LAOS	Asia	5 924 000
LATVIA	Europe	2 307 000
LEBANON	Asia	3 577 000
LESOTHO	Africa	1 795 000
LIBERIA	Africa	3 283 000
LIBYA	Africa	5 853 000
LIECHTENSTEIN	Europe	35 000
LITHUANIA	Europe	3 431 000
LUXEMBOURG	Europe	465 000
MACEDONIA (F.Y.R.O.M.)	Europe	2 034 000
MADAGASCAR	Africa	18 606 000
MALAWI	Africa	12 884 000
MALAYSIA	Asia	25 347 000
MALDIVES	Asia	329 000
MALI	Africa	13 518 000

Capital	Area (sq km)	Area (sq miles)
Tōkyō	377 727	145 841
'Ammān	89 206	34 443
Astana	2 717 300	1 049 155
Nairobi	582 646	224 961
Bairiki	717	277
Kuwait	17 818	6 880
Bishkek	198 500	76 641
Vientiane	236 800	91 429
Rīga	63 700	24 595
Beirut	10 452	4 036
Maseru	30 355	11 720
Monrovia	111 369	43 000
Tripoli	1 759 540	679 362
Vaduz	160	62
Vilnius	65 200	25 174
Luxembourg	2 586	998
Skopje	25 713	9 928
Antananarivo	587 041	226 658
Lilongwe	118 484	45 747
Kuala Lumpur/Putrajaya	332 965	128 559
Male	298	115
Bamako	1 240 140	478 821

	Continent	Population
MALTA	Europe	402 000
MARSHALL ISLANDS	Oceania	62 000
MAURITANIA	Africa	3 069 000
MAURITIUS	Africa	1 245 000
MEXICO	North America	107 029 000
MICRONESIA, FEDERATED STATES OF	Oceania	110 000
MOLDOVA	Europe	4 206 000
MONACO	Europe	35 000
MONGOLIA	Asia	2 646 000
MOROCCO	Africa	31 478 000
MOZAMBIQUE	Africa	19 792 000
MYANMAR (BURMA)	Asia	50 519 000
NAMIBIA	Africa	2 031 000
NAURU	Oceania	14 000
NEPAL	Asia	27 133 000
NETHERLANDS	Europe	16 299 000
NEW ZEALAND	Oceania	4 028 000
NICARAGUA	North America	5 487 000
NIGER	Africa	13 957 000
NIGERIA	Africa	131 530 000
NORTH KOREA	Asia	22 488 000
NORWAY	Europe	4 620 000

Capital	Area (sq km)	Area (sq miles)
Valletta	316	122
Delap-Uliga-Djarrit	181	70
Nouakchott	1 030 700	397 955
Port Louis	2 040	788
Mexico City	1 972 545	761 604
Palikir	701	271
Chişinău	33 700	13 012
Monaco-Ville	2	1
Ulan Bator	1 565 000	604 250
Rabat	446 550	172 414
Maputo	799 380	308 642
Rangoon	676 577	261 228
Windhoek	824 292	318 261
Yaren	21	8
Kathmandu	147 181	56 827
Amsterdam/The Hague	41 526	16 033
Wellington	270 534	104 454
Managua	130 000	50 193
Niamey	1 267 000	489 191
Abuja	923 768	356 669
P'yŏngyang	120 538	46 540
Oslo	323 878	125 050

	Continent	Population
OMAN	Asia	2 567 000
PAKISTAN	Asia	157 935 000
PALAU	Asia	20 000
PANAMA	North America	3 232 000
PAPUA NEW GUINEA	Oceania	5 887 000
PARAGUAY	South America	6 158 000
PERU	South America	27 968 000
PHILIPPINES	Asia	83 054 000
POLAND	Europe	38 530 000
PORTUGAL	Europe	10 495 000
QATAR	Asia	813 000
ROMANIA	Europe	21 711 000
RUSSIAN FEDERATION	Asia/Europe	143 202 000
RWANDA	Africa	9 038 000
ST KITTS AND NEVIS	North America	43 000
ST LUCIA	North America	161 000
ST VINCENT AND THE GRENADINES	North America	119 000
SAMOA	Oceania	185 000
SAN MARINO	Europe	28 000
SÃO TOMÉ AND PRÍNCIPE	Africa	157 000
SAUDI ARABIA	Asia	24 573 000
SENEGAL	Africa	11 658 000

Capital	Area (sq km)	Area (sq miles)
Muscat	309 500	119 499
Islamabad	803 940	310 403
Koror	497	192
Panama City	77 082	29 762
Port Moresby	462 840	178 704
Asunción	406 752	157 048
Lima	1 285 216	496 225
Manila	300 000	115 831
Warsaw	312 683	120 728
Lisbon	88 940	34 340
Doha	11 437	4 416
Bucharest	237 500	91 699
Moscow	17 075 400	6 592 849
Kigali	26 338	10 169
Basseterre	261	101
Castries	616	238
Kingstown	389	150
Apia	2 831	1 093
San Marino	61	24
São Tomé	964	372
Riyadh	2 200 000	849 425
Dakar	196 720	75 954

	Continent	Population
SERBIA AND MONTENEGRO	Europe	10 503 000
SEYCHELLES	Africa	81 000
SIERRA LEONE	Africa	5 525 000
SINGAPORE	Asia	4 326 000
SLOVAKIA	Europe	5 401 000
SLOVENIA	Europe	1 967 000
SOLOMON ISLANDS	Oceania	478 000
SOMALIA	Africa	8 228 000
SOUTH AFRICA, REPUBLIC OF	Africa	47 432 000
SOUTH KOREA	Asia	47 817 000
SPAIN	Europe	43 064 000
SRI LANKA	Asia	20 743 000
SUDAN	Africa	36 233 000
SURINAME	South America	449 000
SWAZILAND	Africa	1 032 000
SWEDEN	Europe	9 041 000
SWITZERLAND	Europe	7 252 000
SYRIA	Asia	19 043 000
TAIWAN	Asia	22 858 000
TAJIKISTAN	Asia	6 507 000
TANZANIA	Africa	38 329 000
THAILAND	Asia	64 233 000

Capital	Area (sq km)	Area (sq miles)
Belgrade	102 173	39 449
Victoria	455	176
Freetown	71 740	27 699
Singapore	639	247
Bratislava	49 035	18 933
Ljubljana	20 251	7 819
Honiara	28 370	10 954
Mogadishu	637 657	246 201
Pretoria (Tshwane)/Cape Town	1 219 090	470 693
Seoul	99 274	38 330
Madrid	504 782	194 897
Sri Jayewardenepura Kotte	65 610	25 332
Khartoum	2 505 813	967 500
Paramaribo	163 820	63 251
Mbabane	17 364	6 704
Stockholm	449 964	173 732
Bern	41 293	15 943
Damascus	185 180	71 498
T'aipei	36 179	13 969
Dushanbe	143 100	55 251
Dodoma	945 087	364 900
Bangkok	513 115	198 115

	Continent	Population
TOGO	Africa	6 145 000
TONGA	Oceania	102 000
TRINIDAD AND TOBAGO	North America	1 305 000
TUNISIA	Africa	10 102 000
TURKEY	Asia	73 193 000
TURKMENISTAN	Asia	4 833 000
TUVALU	Oceania	10 000
UGANDA	Africa	28 816 000
UKRAINE	Europe	46 481 000
UNITED ARAB EMIRATES	Asia	4 496 000
UNITED KINGDOM	Europe	59 668 000
UNITED STATES OF AMERICA	North America	298 213 000
URUGUAY	South America	3 463 000
UZBEKISTAN	Asia	26 593 000
VANUATU	Oceania	211 000
VATICAN CITY	Europe	552
VENEZUELA	South America	26 749 000
VIETNAM	Asia	84 238 000
YEMEN	Asia	20 975 000
ZAMBIA	Africa	11 668 000
ZIMBABWE	Africa	13 010 000

Capital	Area (sq km)	Area (sq miles)
Lomé	56 785	21 925
Nuku'alofa	748	289
Port of Spain	5 130	1 981
Tunis	164 150	63 379
Ankara	779 452	300 948
Aşgabat	488 100	188 456
Vaiaku	25	10
Kampala	241 038	93 065
Kiev	603 700	233 090
Abu Dhabi	77 700	30 000
London	243 609	94 058
Washington	9 826 635	3 794 085
Montevideo	176 215	68 037
Toshkent	447 400	172 742
Port Vila	12 190	4 707
Vatican City	0.5	0.2
Caracas	912 050	352 144
Ha Nôi	329 565	127 246
Şan'ā'	527 968	203 850
Lusaka	752 614	290 586
Harare	390 759	150 873

OVERSEAS AND DISPUTED TERRITORIES

	Continent	Status
American Samoa	Oceania	US Unincorporated Territory
Anguilla	North America	UK Overseas Territory
Aruba	North America	Self-governing Netherlands territory
Azores	Europe	Autonomous Region of Portugal
Bermuda	North America	UK Overseas Territory
Canary Islands	Africa	Autonomous Community of Spain
Cayman Islands	North America	UK Overseas Territory
Ceuta	Africa	Autonomous Community of Spain
Christmas Island	Asia	Australian External Territory
Cocos Islands	Asia	Australian External Territory
Cook Islands	Oceania	Self-governing New Zealand overseas territory
Falkland Islands	South America	UK Overseas Territory
Faroe Islands	Europe	Self-governing Danish territory
French Guiana	South America	French Overseas Department
French Polynesia	Oceania	French Overseas Country
Gaza	Asia	Semi-autonomous region
Gibraltar	Europe	UK Overseas Territory
Greenland	North America	Self-governing Danish territory
Guadeloupe	North America	French Overseas Department
Guam	Oceania	US Unincorporated Territory
Guernsey	Europe	UK Crown Dependency
Isle of Man	Europe	UK Crown Dependency
Jammu and Kashmir	Asia	Disputed territory (India/Pakistan)
Jersey	Europe	UK Crown Dependency

Population	Capital	Area (sq km)	Area (sq miles)
65 000	Fagatogo	197	76
12 000	The Valley	155	60
99 000	Oranjestad	193	75
241 762	Ponta Delgada	2 300	888
64 000	Hamilton	54	21
1 944 700	Santa Cruz de Tenerife/Las Palmas	7 447	2 875
45 000	George Town	259	100
74 931	Ceuta	19	7
1 508	The Settlement	135	52
621	West Island	14	5
18 000	Avarua	293	113
3 000	Stanley	12 170	4 699
47 000	Tórshavn	1 399	540
187 000	Cayenne	90 000	34 749
257 000	Papeete	3 265	1 261
1 406 423	Gaza	363	140
28 000	Gibraltar	7	3
57 000	Nuuk	2 175 600	840 004
448 000	Basse-Terre	1 780	687
170 000	Hågatña	541	209
62 692	St Peter Port	78	30
77 000	Douglas	572	221
13 000 000	Srinagar	222 236	85 806
87 500	St Helier	116	45

	Continent	Status
Madeira	Africa	Autonomous Region of Portugal
Martinique	North America	French Overseas Department
Mayotte	Africa	French Departmental Collectivity
Melilla	Africa	Autonomous Community of Spain
Montserrat	North America	UK Overseas Territory
Netherlands Antilles	North America	Self-governing Netherlands territory
New Caledonia	Oceania	French Overseas Country
Niue	Oceania	Self-governing New Zealand overseas territory
Norfolk Island	Oceania	Australian External Territory
Northern Mariana Islands	Oceania	US Commonwealth
Pitcairn Islands	Oceania	UK Overseas Territory
Puerto Rico	North America	US Commonwealth
Réunion	Africa	French Overseas Department
St Helena	Africa	UK Overseas Territory
St Pierre and Miquelon	North America	French Territorial Collectivity
Tokelau	Oceania	New Zealand Overseas Territory
Turks and Caicos Islands	North America	UK Overseas Territory
Virgin Islands (U.K.)	North America	UK Overseas Territory
Virgin Islands (U.S.)	North America	US Unincorporated Territory
Wallis and Futuna Islands	Oceania	French Overseas Territory
West Bank	Asia	Disputed territory
Western Sahara	Africa	Disputed territory (Morocco)

Population	Capital	Area (sq km)	Area (sq miles)
245 012	Funchal	779	301
396 000	Fort-de-France	1 079	417
186 026	Dzaoudzi	373	144
68 463	Melilla	13	5
4 000	Plymouth	100	39
183 000	Willemstad	800	309
237 000	Nouméa	19 058	7 358
1 000	Alofi	258	100
2 601	Kingston	35	14
81 000	Capitol Hill	477	184
47	Adamstown	45	17
3 955 000	San Juan	9 104	3 515
785 000	St-Denis	2 551	985
5 000	Jamestown	121	47
6 000	St-Pierre	242	93
1 000	-	10	4
26 000	Grand Turk	430	166
22 000	Road Town	153	59
112 000	Charlotte Amalie	352	136
15 000	Matā'utu	274	106
2 421 491	-	5 860	2 263
341 000	Laâyoune	266 000	102 703

LANGUAGES, RELIGIONS AND CURRENCIES

	Main languages
AFGHANISTAN	Dari, Pushtu, Uzbek, Turkmen
ALBANIA	Albanian, Greek
ALGERIA	Arabic, French, Berber
American Samoa	Samoan, English
ANDORRA	Spanish, Catalan, French
ANGOLA	Portuguese, Bantu, local languages
Anguilla	English
ANTIGUA AND BARBUDA	English, creole
ARGENTINA	Spanish, Italian, Amerindian languages
ARMENIA	Armenian, Azeri
Aruba	Papiamento, Dutch, English
AUSTRALIA	English, Italian, Greek
AUSTRIA	German, Croatian, Turkish
AZERBAIJAN	Azeri, Armenian, Russian, Lezgian
Azores	Portuguese
THE BAHAMAS	English, creole
BAHRAIN	Arabic, English
BANGLADESH	Bengali, English
BARBADOS	English, creole
BELARUS	Belorussian, Russian

Main religions	Currency
Sunni Muslim, Shi'a Muslim	Afghani
Sunni Muslim, Albanian Orthodox, Roman Catholic	Lek
Sunni Muslim	Algerian dinar
Protestant, Roman Catholic	US dollar
Roman Catholic	Euro
Roman Catholic, Protestant, traditional beliefs	Kwanza
Protestant, Roman Catholic	East Caribbean dollar
Protestant, Roman Catholic	East Caribbean dollar
Roman Catholic, Protestant	Argentinian peso
Armenian Orthodox	Dram
Roman Catholic, Protestant	Aruban florin
Protestant, Roman Catholic, Orthodox	Australian dollar
Roman Catholic, Protestant	Euro
Shi'a Muslim, Sunni Muslim, Russian and Armenian Orthodox	Azerbaijani manat
Roman Catholic, Protestant	Euro
Protestant, Roman Catholic	Bahamian dollar
Shi'a Muslim, Sunni Muslim, Christian	Bahraini dinar
Sunni Muslim, Hindu	Taka
Protestant, Roman Catholic	Barbados dollar
Belorussian Orthodox, Roman Catholic	Belarus rouble

	Main languages
BELGIUM	Dutch (Flemish), French (Walloon), German
BELIZE	English, Spanish, Mayan, creole
BENIN	French, Fon, Yoruba, Adja, local languages
Bermuda	English
BHUTAN	Dzongkha, Nepali, Assamese
BOLIVIA	Spanish, Quechua, Aymara
BOSNIA-HERZEGOVINA	Bosnian, Serbian, Croatian
BOTSWANA	English, Setswana, Shona, local languages
BRAZIL	Portuguese
BRUNEI	Malay, English, Chinese
BULGARIA	Bulgarian, Turkish, Romany, Macedonian
BURKINA	French, Moore (Mossi), Fulani, local languages
BURUNDI	Kirundi (Hutu, Tutsi), French
CAMBODIA	Khmer, Vietnamese
CAMEROON	French, English, Fang, Bamileke, local languages
CANADA	English, French, local languages
Canary Islands	Spanish
CAPE VERDE	Portuguese, creole
Cayman Islands	English
CENTRAL AFRICAN REPUBLIC	French, Sango, Banda, Baya, local languages
Ceuta	Spanish, Arabic

Main religions	Currency
Roman Catholic, Protestant	Euro
Roman Catholic, Protestant	Belize dollar
Traditional beliefs, Roman Catholic, Sunni Muslim	CFA franc
Protestant, Roman Catholic	Bermuda dollar
Buddhist, Hindu	Ngultrum, Indian rupee
Roman Catholic, Protestant, Baha'i	Boliviano
Sunni Muslim, Serbian Orthodox, Roman Catholic, Protestant	Marka
Traditional beliefs, Protestant, Roman Catholic	Pula
Roman Catholic, Protestant	Real
Sunni Muslim, Buddhist, Christian	Brunei dollar
Bulgarian Orthodox, Sunni Muslim	Lev
Sunni Muslim, traditional beliefs, Roman Catholic	CFA franc
Roman Catholic, traditional beliefs, Protestant	Burundian franc
Buddhist, Roman Catholic, Sunni Muslim	Riel
Roman Catholic, traditional beliefs, Sunni Muslim, Protestant	CFA franc
Roman Catholic, Protestant, Eastern Orthodox, Jewish	Canadian dollar
Roman Catholic	Euro
Roman Catholic, Protestant	Cape Verde escudo
Protestant, Roman Catholic	Cayman Islands dollar
Protestant, Roman Catholic, traditional beliefs, Sunni Muslim	CFA franc
Roman Catholic, Muslim	Euro

	Main languages
CHAD	Arabic, French, Sara, local languages
CHILE	Spanish, Amerindian languages
CHINA	Mandarin, Wu, Cantonese, Hsiang, regional languages
Christmas Island	English
Cocos Islands	English
COLOMBIA	Spanish, Amerindian languages
COMOROS	Comorian, French, Arabic
CONGO	French, Kongo, Monokutuba, local languages
CONGO, DEMOCRATIC REPUBLIC OF THE	French, Lingala, Swahili, Kongo, local languages
Cook Islands	English, Maori
COSTA RICA	Spanish
CÔTE D'IVOIRE	French, creole, Akan, local languages
CROATIA	Croatian, Serbian
CUBA	Spanish
CYPRUS	Greek, Turkish, English
CZECH REPUBLIC	Czech, Moravian, Slovak
DENMARK	Danish
DJIBOUTI	Somali, Afar, French, Arabic
DOMINICA	English, creole
DOMINICAN REPUBLIC	Spanish, creole

Main religions	Currency
Sunni Muslim, Roman Catholic, Protestant, traditional beliefs	CFA franc
Roman Catholic, Protestant	Chilean peso
Confucian, Taoist, Buddhist, Christian, Sunni Muslim	Yuan, Hong Kong dollar, Macao pataca
Buddhist, Sunni Muslim, Protestant, Roman Catholic	Australian dollar
Sunni Muslim, Christian	Australian dollar
Roman Catholic, Protestant	Colombian peso
Sunni Muslim, Roman Catholic	Comoros franc
Roman Catholic, Protestant, traditional beliefs, Sunni Muslim	CFA franc
Christian, Sunni Muslim	Congolese franc
Protestant, Roman Catholic	New Zealand dollar
Roman Catholic, Protestant	Costa Rican colón
Sunni Muslim, Roman Catholic, traditional beliefs, Protestant	CFA franc
Roman Catholic, Serbian Orthodox, Sunni Muslim	Kuna
Roman Catholic, Protestant	Cuban peso
Greek Orthodox, Sunni Muslim	Cyprus pound
Roman Catholic, Protestant	Czech koruna
Protestant	Danish krone
Sunni Muslim, Christian	Djibouti franc
Roman Catholic, Protestant	East Caribbean dollar
Roman Catholic, Protestant	Dominican peso

	Main languages
EAST TIMOR	Portuguese, Tetun, English
ECUADOR	Spanish, Quechua, and other Amerindian languages
EGYPT	Arabic
EL SALVADOR	Spanish
EQUATORIAL GUINEA	Spanish, French, Fang
ERITREA	Tigrinya, Tigre
ESTONIA	Estonian, Russian
ETHIOPIA	Oromo, Amharic, Tigrinya, local languages
Falkland Islands	English
Faroe Islands	Faroese, Danish
FIJI	English, Fijian, Hindi
FINLAND	Finnish, Swedish
FRANCE	French, Arabic
French Guiana	French, creole
French Polynesia	French, Tahitian, Polynesian languages
GABON	French, Fang, local languages
THE GAMBIA	English, Malinke, Fulani, Wolof
Gaza	Arabic
GEORGIA	Georgian, Russian, Armenian, Azeri, Ossetian, Abkhaz
GERMANY	German, Turkish
GHANA	English, Hausa, Akan, local languages
Gibraltar	English, Spanish

Main religions	Currency
Roman Catholic	US dollar
Roman Catholic	US dollar
Sunni Muslim, Coptic Christian	Egyptian pound
Roman Catholic, Protestant	El Salvador colón, US dollar
Roman Catholic, traditional beliefs	CFA franc
Sunni Muslim, Coptic Christian	Nakfa
Protestant, Estonian and Russian Orthodox	Kroon
Ethiopian Orthodox, Sunni Muslim, traditional beliefs	Birr
Protestant, Roman Catholic	Falkland Islands pound
Protestant	Danish krone
Christian, Hindu, Sunni Muslim	Fiji dollar
Protestant, Greek Orthodox	Euro
Roman Catholic, Protestant, Sunni Muslim	Euro
Roman Catholic	Euro
Protestant, Roman Catholic	CFP franc
Roman Catholic, Protestant, traditional beliefs	CFA franc
Sunni Muslim, Protestant	Dalasi
Sunni Muslim, Shi'a Muslim	Israeli shekel
Georgian Orthodox, Russian Orthodox, Sunni Muslim	Lari
Protestant, Roman Catholic	Euro
Christian, Sunni Muslim, traditional beliefs	Cedi
Roman Catholic, Protestant, Sunni Muslim	Gibraltar pound

	Main languages
GREECE	Greek
Greenland	Greenlandic, Danish
GRENADA	English, creole
Guadeloupe	French, creole
Guam	Chamorro, English, Tapalog
GUATEMALA	Spanish, Mayan languages
Guernsey	English, French
GUINEA	French, Fulani, Malinke, local languages
GUINEA-BISSAU	Portuguese, crioulo, local languages
GUYANA	English, creole, Amerindian languages
HAITI	French, creole
HONDURAS	Spanish, Amerindian languages
HUNGARY	Hungarian
ICELAND	Icelandic
INDIA	Hindi, English, many regional languages
INDONESIA	Indonesian, local languages
IRAN	Farsi, Azeri, Kurdish, regional languages
IRAQ	Arabic, Kurdish, Turkmen
IRELAND	English, Irish
Isle of Man	English
ISRAEL	Hebrew, Arabic
ITALY	Italian
JAMAICA	English, creole
JAPAN	Japanese

Main religions	Currency
Greek Orthodox, Sunni Muslim	Euro
Protestant	Danish krone
Roman Catholic, Protestant	East Caribbean dollar
Roman Catholic	Euro
Roman Catholic	US dollar
Roman Catholic, Protestant	Quetzal, US dollar
Protestant, Roman Catholic	Pound sterling
Sunni Muslim, traditional beliefs, Christian	Guinea franc
Traditional beliefs, Sunni Muslim, Christian	CFA franc
Protestant, Hindu, Roman Catholic, Sunni Muslim	Guyana dollar
Roman Catholic, Protestant, Voodoo	Gourde
Roman Catholic, Protestant	Lempira
Roman Catholic, Protestant	Forint
Protestant	Icelandic króna
Hindu, Sunni Muslim, Shi'a Muslim, Sikh, Christian	Indian rupee
Sunni Muslim, Protestant, Roman Catholic, Hindu, Buddhist	Rupiah
Shi'a Muslim, Sunni Muslim	Iranian rial
Shi'a Muslim, Sunni Muslim, Christian	Iraqi dinar
Roman Catholic, Protestant	Euro
Protestant, Roman Catholic	Pound sterling
Jewish, Sunni Muslim, Christian, Druze	Shekel
Roman Catholic	Euro
Protestant, Roman Catholic	Jamaican dollar
Shintoist, Buddhist, Christian	Yen

	Main languages
Jersey	English, French
JORDAN	Arabic
KAZAKHSTAN	Kazakh, Russian, Ukrainian, German, Uzbek, Tatar
KENYA	Swahili, English, local languages
KIRIBATI	Gilbertese, English
KUWAIT	Arabic
KYRGYZSTAN	Kyrgyz, Russian, Uzbek
LAOS	Lao, local languages
LATVIA	Latvian, Russian
LEBANON	Arabic, Armenian, French
LESOTHO	Sesotho, English, Zulu
LIBERIA	English, creole, local languages
LIBYA	Arabic, Berber
LIECHTENSTEIN	German
LITHUANIA	Lithuanian, Russian, Polish
LUXEMBOURG	Letzeburgish, German, French
MACEDONIA (F.Y.R.O.M.)	Macedonian, Albanian, Turkish
MADAGASCAR	Malagasy, French
Madeira	Portuguese
MALAWI	Chichewa, English, local languages
MALAYSIA	Malay, English, Chinese, Tamil, local languages
MALDIVES	Divehi (Maldivian)
MALI	French, Bambara, local languages

Main religions	Currency
Protestant, Roman Catholic	Pound sterling
Sunni Muslim, Christian	Jordanian dinar
Sunni Muslim, Russian Orthodox, Protestant	Tenge
Christian, traditional beliefs	Kenyan shilling
Roman Catholic, Protestant	Australian dollar
Sunni Muslim, Shi'a Muslim, Christian, Hindu	Kuwaiti dinar
Sunni Muslim, Russian Orthodox	Kyrgyz som
Buddhist, traditional beliefs	Kip
Protestant, Roman Catholic, Russian Orthodox	Lats
Shi'a Muslim, Sunni Muslim, Christian	Lebanese pound
Christian, traditional beliefs	Loti, S. African rand
Traditional beliefs, Christian, Sunni Muslim	Liberian dollar
Sunni Muslim	Libyan dinar
Roman Catholic, Protestant	Swiss franc
Roman Catholic, Protestant, Russian Orthodox	Litas
Roman Catholic	Euro
Macedonian Orthodox, Sunni Muslim	Macedonian denar
Traditional beliefs, Christian, Sunni Muslim	Malagasy Ariary, Malagasy franc
Roman Catholic, Protestant	Euro
Christian, traditional beliefs, Sunni Muslim	Malawian kwacha
Sunni Muslim, Buddhist, Hindu, Christian, traditional beliefs	Ringgit
Sunni Muslim	Rufiyaa
Sunni Muslim, traditional beliefs, Christian	CFA franc

	Main languages
MALTA	Maltese, English
MARSHALL ISLANDS	English, Marshallese
Martinique	French, creole
MAURITANIA	Arabic, French, local languages
MAURITIUS	English, creole, Hindi, Bhojpuri, French
Mayotte	French, Mahorian
Melilla	Spanish, Arabic
MEXICO	Spanish, Amerindian languages
MICRONESIA, FEDERATED STATES OF	English, Chuukese, Pohnpeian, local languages
MOLDOVA	Romanian, Ukrainian, Gagauz, Russian,
MONACO	French, Monegasque, Italian
MONGOLIA	Khalka (Mongolian), Kazakh, local languages
Montserrat	English
MOROCCO	Arabic, Berber, French
MOZAMBIQUE	Portuguese, Makua, Tsonga, local languages
MYANMAR (BURMA)	Burmese, Shan, Karen, local languages
NAMIBIA	English, Afrikaans, German, Ovambo, local languages
NAURU	Nauruan, English
NEPAL	Nepali, Maithili, Bhojpuri, English, local languages
NETHERLANDS	Dutch, Frisian
Netherlands Antilles	Dutch, Papiamento, English
New Caledonia	French, local languages

Main religions	Currency
Roman Catholic	Maltese lira
Protestant, Roman Catholic	US dollar
Roman Catholic, traditional beliefs	Euro
Sunni Muslim	Ouguiya
Hindu, Roman Catholic, Sunni Muslim	Mauritius rupee
Sunni Muslim, Christian	Euro
Roman Catholic, Muslim	Euro
Roman Catholic, Protestant	Mexican peso
Roman Catholic, Protestant	US dollar
Romanian Orthodox, Russian Orthodox	Moldovan leu
Roman Catholic	Euro
Buddhist, Sunni Muslim	Tugrik (tögrög)
Protestant, Roman Catholic	East Caribbean dollar
Sunni Muslim	Moroccan dirham
Traditional beliefs, Roman Catholic, Sunni Muslim	Metical
Buddhist, Christian, Sunni Muslim	Kyat
Protestant, Roman Catholic	Namibian dollar
Protestant, Roman Catholic	Australian dollar
Hindu, Buddhist, Sunni Muslim	Nepalese rupee
Roman Catholic, Protestant, Sunni Muslim	Euro
Roman Catholic, Protestant	Netherlands Antilles guilder
Roman Catholic, Protestant, Sunni Muslim	CFP franc

	Main languages
NEW ZEALAND	English, Maori
NICARAGUA	Spanish, Amerindian languages
NIGER	French, Hausa, Fulani, local languages
NIGERIA	English, Hausa, Yoruba, Ibo, Fulani, local languages
Niue	English, Niuean
Norfolk Island	English
NORTH KOREA	Korean
Northern Mariana Islands	English, Chamorro, local languages
NORWAY	Norwegian
OMAN	Arabic, Baluchi, Indian languages
PAKISTAN	Urdu, Punjabi, Sindhi, Pushtu, English
PALAU	Palauan, English
PANAMA	Spanish, English, Amerindian languages
PAPUA NEW GUINEA	English, Tok Pisin (creole), local languages
PARAGUAY	Spanish, Guarani
PERU	Spanish, Quechua, Aymara
PHILIPPINES	English, Filipino, Tagalog, Cebuano, local languages
Pitcairn Islands	English
POLAND	Polish, German
PORTUGAL	Portuguese
Puerto Rico	Spanish, English
QATAR	Arabic

Main religions	Currency
Protestant, Roman Catholic	New Zealand dollar
Roman Catholic, Protestant	Córdoba
Sunni Muslim, traditional beliefs	CFA franc
Sunni Muslim, Christian, traditional beliefs	Naira
Christian	New Zealand dollar
Protestant, Roman Catholic	Australian dollar
Traditional beliefs, Chondoist, Buddhist	North Korean won
Roman Catholic	US dollar
Protestant, Roman Catholic	Norwegian krone
Ibadhi Muslim, Sunni Muslim	Omani riyal
Sunni Muslim, Shi'a Muslim, Christian, Hindu	Pakistani rupee
Roman Catholic, Protestant, traditional beliefs	US dollar
Roman Catholic, Protestant, Sunni Muslim	Balboa
Protestant, Roman Catholic, traditional beliefs	Kina
Roman Catholic, Protestant	Guaraní
Roman Catholic, Protestant	Sol
Roman Catholic, Protestant, Sunni Muslim, Aglipayan	Philippine peso
Protestant	New Zealand dollar
Roman Catholic, Polish Orthodox	Złoty
Roman Catholic, Protestant	Euro
Roman Catholic, Protestant	US dollar
Sunni Muslim	Qatari riyal

	Main languages
Réunion	French, creole
ROMANIA	Romanian, Hungarian
RUSSIAN FEDERATION	Russian, Tatar, Ukrainian, local languages
RWANDA	Kinyarwanda, French, English
St Helena	English
ST KITTS AND NEVIS	English, creole
ST LUCIA	English, creole
St Pierre and Miquelon	French
ST VINCENT AND THE GRENADINES	English, creole
SAMOA	Samoan, English
SAN MARINO	Italian
SÃO TOMÉ AND PRÍNCIPE	Portuguese, creole
SAUDI ARABIA	Arabic
SENEGAL	French, Wolof, Fulani, local languages
SERBIA AND MONTENEGRO	Serbian, Albanian, Hungarian
SEYCHELLES	English, French, creole
SIERRA LEONE	English, creole, Mende, Temne, local languages
SINGAPORE	Chinese, English, Malay, Tamil
SLOVAKIA	Slovak, Hungarian, Czech
SLOVENIA	Slovene, Croatian, Serbian
SOLOMON ISLANDS	English, creole, local languages

Main religions	Currency
Roman Catholic	Euro
Romanian Orthodox, Protestant, Roman Catholic	Romanian leu
Russian Orthodox, Sunni Muslim, Protestant	Russian rouble
Roman Catholic, traditional beliefs, Protestant	Rwandan franc
Protestant, Roman Catholic	St Helena pound
Protestant, Roman Catholic	East Caribbean dollar
Roman Catholic, Protestant	East Caribbean dollar
Roman Catholic	Euro
Protestant, Roman Catholic	East Caribbean dollar
Protestant, Roman Catholic	Tala
Roman Catholic	Euro
Roman Catholic, Protestant	Dobra
Sunni Muslim, Shi'a Muslim	Saudi Arabian riyal
Sunni Muslim, Roman Catholic, traditional beliefs	CFA franc
Serbian Orthodox, Montenegrin Orthodox, Sunni Muslim	Serbian dinar, Euro
Roman Catholic, Protestant	Seychelles rupee
Sunni Muslim, traditional beliefs	Leone
Buddhist, Taoist, Sunni Muslim, Christian, Hindu	Singapore dollar
Roman Catholic, Protestant, Orthodox	Slovakian koruna
Roman Catholic, Protestant	Tólar
Protestant, Roman Catholic	Solomon Islands dollar

	Main languages
SOMALIA	Somali, Arabic
SOUTH AFRICA, REPUBLIC OF	Afrikaans, English, nine official local languages
SOUTH KOREA	Korean
SPAIN	Castilian, Catalan, Galician, Basque
SRI LANKA	Sinhalese, Tamil, English
SUDAN	Arabic, Dinka, Nubian, Beja, Nuer, local languages
SURINAME	Dutch, Surinamese, English, Hindi,
SWAZILAND	Swazi, English
SWEDEN	Swedish
SWITZERLAND	German, French, Italian, Romansch
SYRIA	Arabic, Kurdish, Armenian
TAIWAN	Mandarin, Min, Hakka, local languages
TAJIKISTAN	Tajik, Uzbek, Russian
TANZANIA	Swahili, English, Nyamwezi, local languages
THAILAND	Thai, Lao, Chinese, Malay, Mon-Khmer languages
TOGO	French, Ewe, Kabre, local languages
Tokelau	English, Tokelauan
TONGA	Tongan, English
TRINIDAD AND TOBAGO	English, creole, Hindi
TUNISIA	Arabic, French
TURKEY	Turkish, Kurdish

Main religions	Currency
Sunni Muslim	Somali shilling
Protestant, Roman Catholic, Sunni Muslim, Hindu	Rand
Buddhist, Protestant, Roman Catholic	South Korean won
Roman Catholic	Euro
Buddhist, Hindu, Sunni Muslim, Roman Catholic	Sri Lankan rupee
Sunni Muslim, traditional beliefs, Christian	Sudanese dinar
Hindu, Roman Catholic, Protestant, Sunni Muslim	Suriname guilder
Christian, traditional beliefs	Emalangeni, South African rand
Protestant, Roman Catholic	Swedish krona
Roman Catholic, Protestant	Swiss franc
Sunni Muslim, Shi'a Muslim, Christian	Syrian pound
Buddhist, Taoist, Confucian, Christian	Taiwan dollar
Sunni Muslim	Somoni
Shi'a Muslim, Sunni Muslim, traditional beliefs, Christian	Tanzanian shilling
Buddhist, Sunni Muslim	Baht
Traditional beliefs, Christian, Sunni Muslim	CFA franc
Christian	New Zealand dollar
Protestant, Roman Catholic	Pa'anga
Roman Catholic, Hindu, Protestant, Sunni Muslim	Trinidad and Tobago dollar
Sunni Muslim	Tunisian dinar
Sunni Muslim, Shi'a Muslim	Turkish lira

	Main languages
TURKMENISTAN	Turkmen, Uzbek, Russian
Turks and Caicos Islands	English
TUVALU	Tuvaluan, English
UGANDA	English, Swahili, Luganda, local languages
UKRAINE	Ukrainian, Russian
UNITED ARAB EMIRATES	Arabic, English
UNITED KINGDOM	English, Welsh, Gaelic
UNITED STATES OF AMERICA	English, Spanish
URUGUAY	Spanish
UZBEKISTAN	Uzbek, Russian, Tajik, Kazakh
VANUATU	English, Bislama (creole), French
VATICAN CITY	Italian
VENEZUELA	Spanish, Amerindian languages
VIETNAM	Vietnamese, Thai, Khmer, Chinese, local languag
Virgin Islands (U.K.)	English
Virgin Islands (U.S.)	English, Spanish
Wallis and Futuna Islands	French, Wallisian, Futunian
West Bank	Arabic, Hebrew
Western Sahara	Arabic
YEMEN	Arabic
ZAMBIA	English, Bemba, Nyanja, Tonga, local languages
ZIMBABWE	English, Shona, Ndebele

Main religions	Currency
Sunni Muslim, Russian Orthodox	Turkmen manat
Protestant	US dollar
Protestant	Australian dollar
Roman Catholic, Protestant, Sunni Muslim, traditional beliefs	Ugandan shilling
Ukrainian Orthodox, Ukrainian Catholic, Roman Catholic	Hryvnia
Sunni Muslim, Shi'a Muslim	UAE dirham
Protestant, Roman Catholic, Muslim	Pound sterling
Protestant, Roman Catholic, Sunni Muslim, Jewish	US dollar
Roman Catholic, Protestant, Jewish	Uruguayan peso
Sunni Muslim, Russian Orthodox	Uzbek som
Protestant, Roman Catholic, traditional beliefs	Vatu
Roman Catholic	Euro
Roman Catholic, Protestant	Bolívar
Buddhist, Taoist, Roman Catholic, Cao Dai, Hoa Hao	Dong
Protestant, Roman Catholic	US dollar
Protestant, Roman Catholic	US dollar
Roman Catholic	CFP franc
Sunni Muslim, Jewish, Shi'a Muslim, Christian	Jordanian dinar, Isreali shekel
Sunni Muslim	Moroccan dirham
Sunni Muslim, Shi'a Muslim	Yemeni riyal
Christian, traditional beliefs	Zambian kwacha
Christian, traditional beliefs	Zimbabwean dollar

TIME ZONES

TIME ZONES AND THE INTERNATIONAL DATE LINE

TIME ZONES

The system of timekeeping throughout the world is based on twenty-four time zones, each stretching over fifteen degrees of longitude – the distance equivalent to a time difference of one hour. The prime, or Greenwich Meridian (0 degrees longitude), is the basis for Greenwich Mean Time (GMT), also known as Universal Coordinated Time (UTC) by which other time zones are measured. This universal reference point was agreed by delegates from twenty-six countries at the International Meridian Conference in Washington, DC in 1884. Prior to this, many separate central meridians were in use, for navigational and reference purposes, including London, Paris, Cadiz and Stockholm.

Many countries, including the UK and USA, use daylight saving time (DST) in order to maximize daylight hours in summer. Clocks are put forward one hour in spring and back one hour in autumn. DST was first introduced to the UK during the First World War to reduce the demand for artificial heating and lighting.

Time zone boundaries can be altered to suit international or internal boundaries. China uses only one time zone although it should theoretically have five, while the Russian Federation stretches over eleven zones. Mainland USA has four time zones – Eastern Standard, Central Standard, Mountain Standard and Pacific Standard Time – which do not always follow state boundaries.

THE INTERNATIONAL DATE LINE

The International Date Line is an imaginary line passing down the Pacific Ocean at approximately 180° west (or east) of Greenwich, across which the date changes by one day. To the left (west) of the line the date is always one day ahead of the right (east). If travelling eastwards across the line, travellers must move their calendars back one day.

The position and status of the line was agreed at the same conference at which Greenwich was adopted as the prime meridian. The line has no international legal status and countries near to it can choose which date they will observe. It was amended most recently so that Caroline Island, in the Pacific nation of Kiribati, would be the first land area to greet the year 2000. The island was renamed Millennium Island in recognition of this.

WEB LINKS

Greenwich Royal Observatory	www.rog.nmm.ac.uk
World time	wwp.greenwichmeantime.com
World time zones	www.worldtimezones.com
Official US time	www.time.gov/
International Date Line	aa.usno.navy.mil/faq/docs/international_date.html

THE WORLD'S TIME ZONES

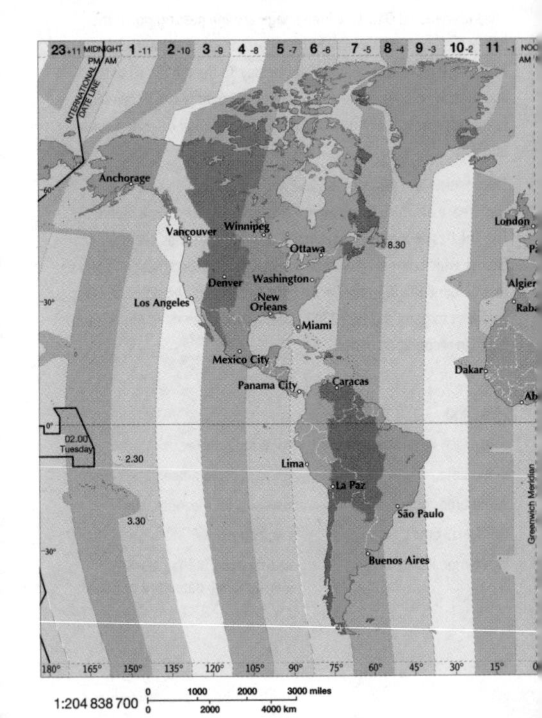

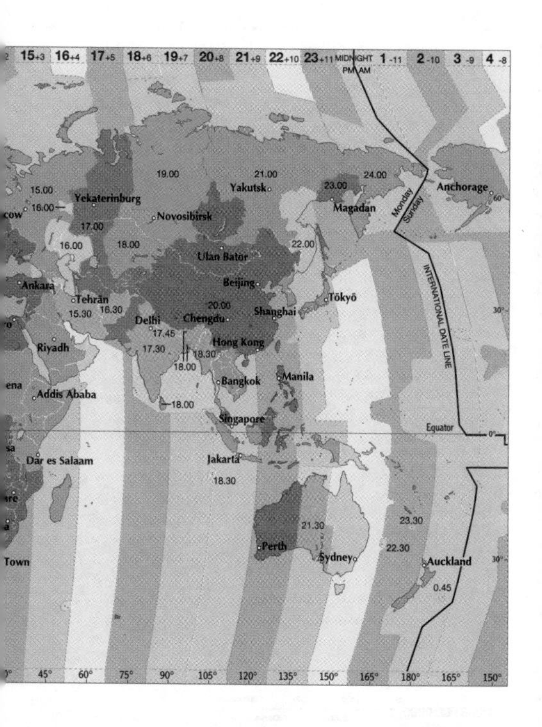
15+3 16+4 17+5 18+6 19+7 20+8 21+9 22+10 23+11 MIDNIGHT 1 -11 2 -10 3 -9 4 -8
PM AM
Yekaterinburg
Novosibirsk
Yakutsk
Magadan
Anchorage
Ulan Bator
Beijing
Ankara
Tehrān
Tōkyō
Shanghai
Delhi
Chengdu
Hong Kong
Riyadh
Bangkok
Manila
Addis Ababa
Singapore
Dar es Salaam
Jakarta
Perth
Sydney
Auckland
Monday
Sunday
INTERNATIONAL DATE LINE
Equator
15.00
16.00
17.00
18.00
19.00
21.00
23.00
24.00
22.00
20.00
15.30
16.30
17.45
17.30
18.30
18.00
21.30
23.30
22.30
0.45
45° 60° 75° 90° 105° 120° 135° 150° 165° 180° 165° 150°

CLIMATE

THE WORLD'S CLIMATIC REGIONS

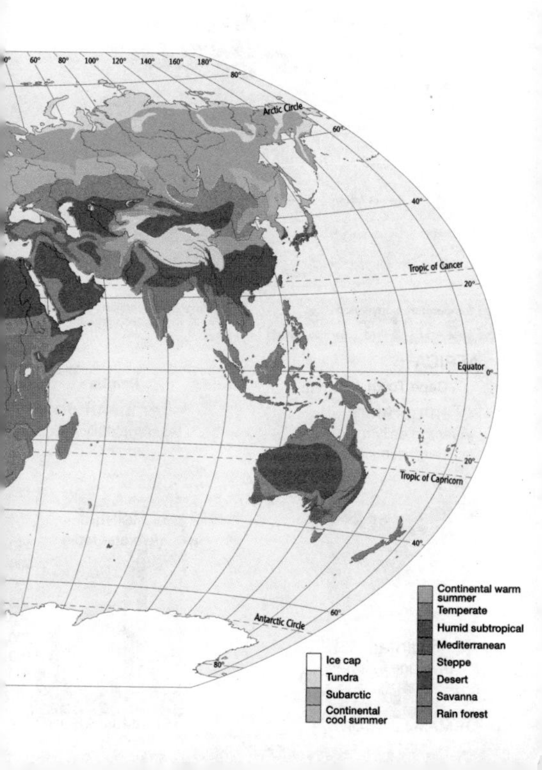
60°
80°
100°
120°
140°
160°
180°
80°
Arctic Circle
60°
40°
Tropic of Cancer
20°
Equator 0°
20°
Tropic of Capricorn
40°
60°
Antarctic Circle
80°
Ice cap
Tundra
Subarctic
Continental cool summer
Continental warm summer
Temperate
Humid subtropical
Mediterranean
Steppe
Desert
Savanna
Rain forest

CLIMATE AROUND THE WORLD

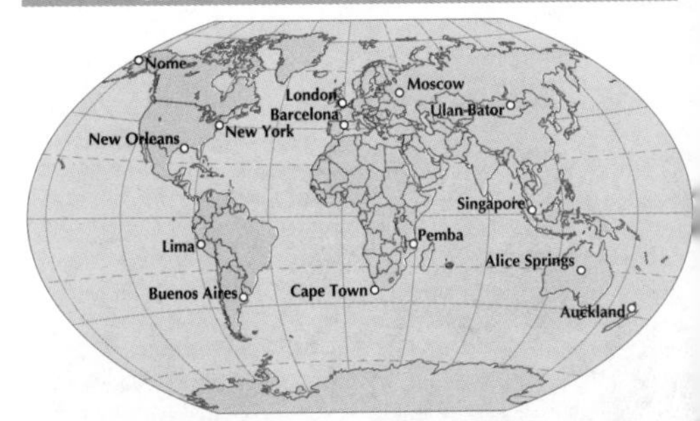

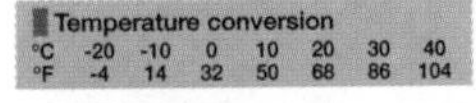

Temperature conversion

°C	-20	-10	0	10	20	30	40
°F	-4	14	32	50	68	86	104

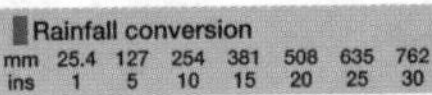

Rainfall conversion

mm	25.4	127	254	381	508	635	762
ins	1	5	10	15	20	25	30

AFRICA

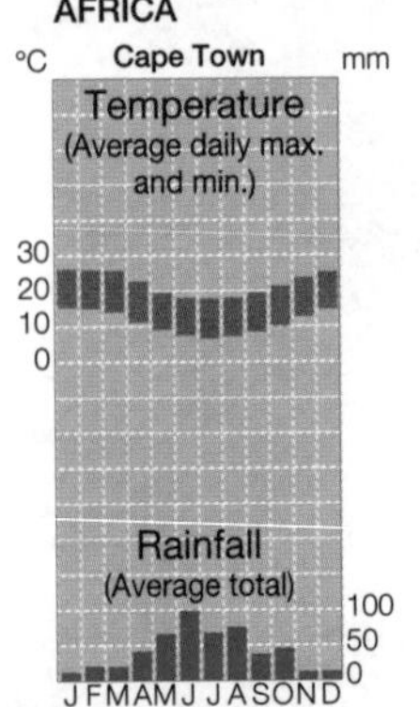

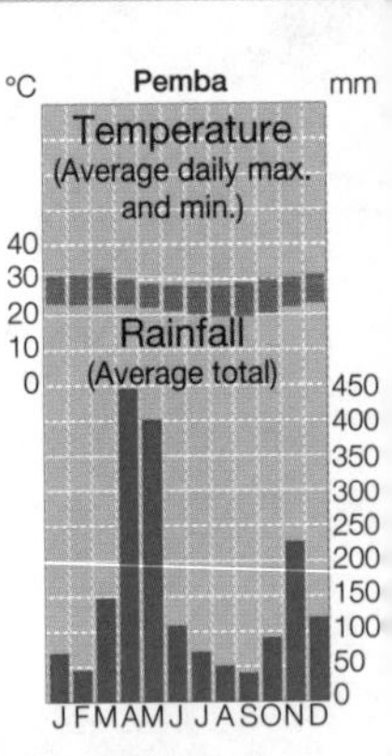

ASIA

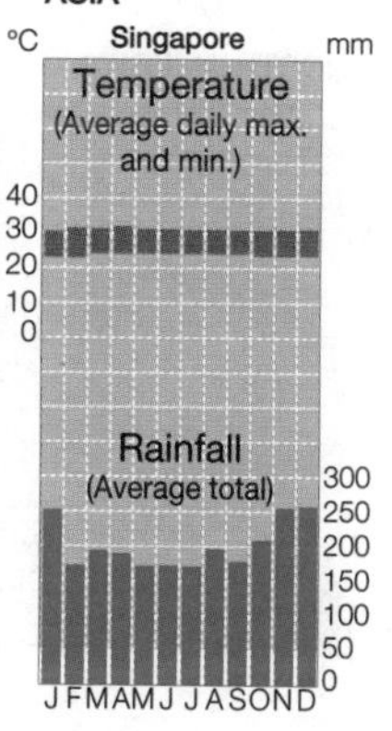
°C
Singapore
mm
Temperature
(Average daily max. and min.)
40
30
20
10
0
Rainfall
(Average total)
300
250
200
150
100
50
0
J F M A M J J A S O N D

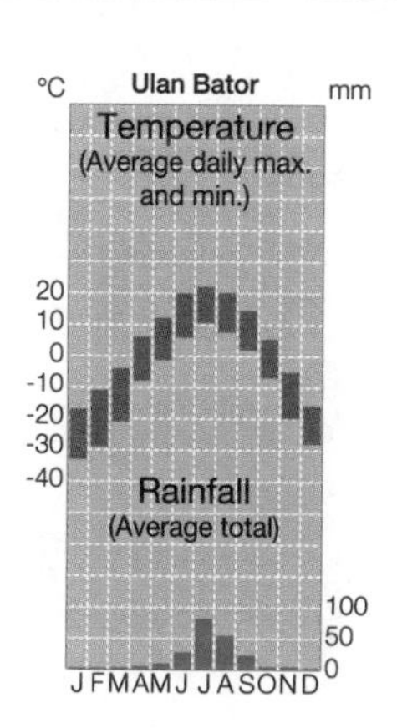
°C
Ulan Bator
mm
Temperature
(Average daily max. and min.)
20
10
0
-10
-20
-30
-40
Rainfall
(Average total)
100
50
0
J F M A M J J A S O N D

EUROPE

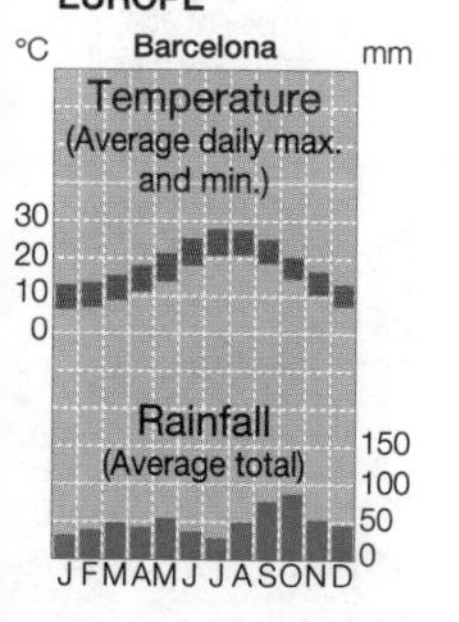
°C
Barcelona
mm
Temperature
(Average daily max. and min.)
30
20
10
0
Rainfall
(Average total)
150
100
50
0
J F M A M J J A S O N D

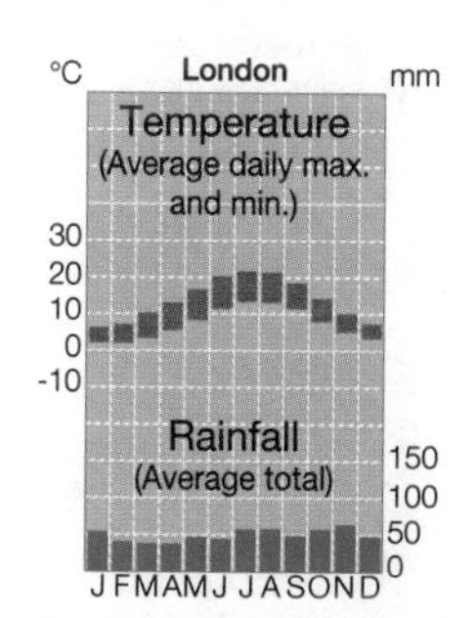
°C
London
mm
Temperature
(Average daily max. and min.)
30
20
10
0
-10
Rainfall
(Average total)
150
100
50
0
J F M A M J J A S O N D

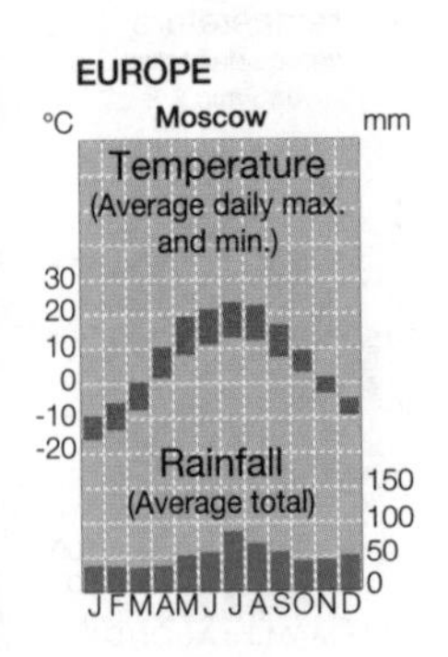
EUROPE
°C
Moscow
mm
Temperature
(Average daily max. and min.)
30
20
10
0
-10
-20
Rainfall
(Average total)
150
100
50
0
J F M A M J J A S O N D

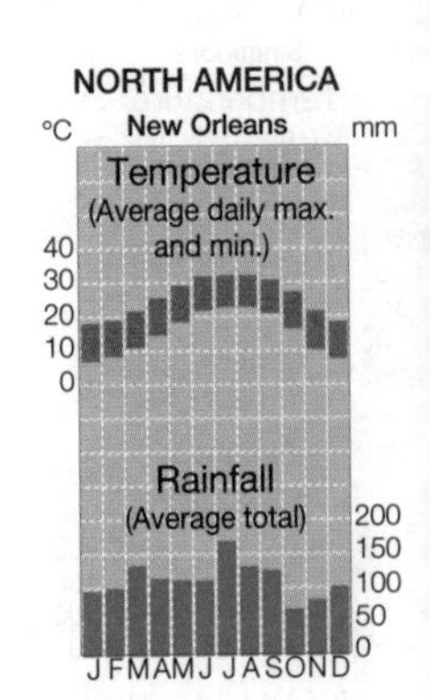
NORTH AMERICA
°C
New Orleans
mm
Temperature
(Average daily max. and min.)
40
30
20
10
0
Rainfall
(Average total)
200
150
100
50
0
J F M A M J J A S O N D

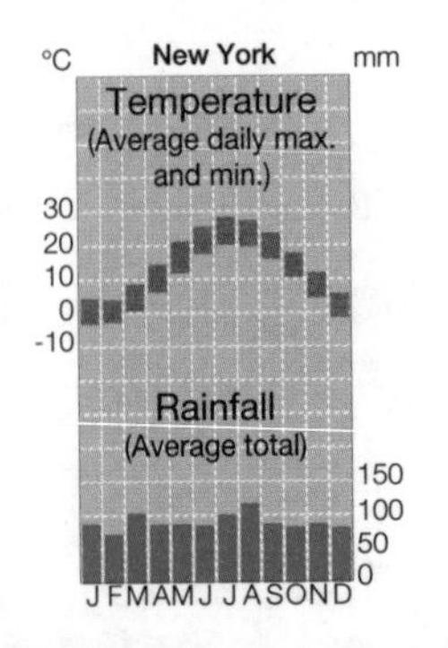
°C
New York
mm
Temperature
(Average daily max. and min.)
30
20
10
0
-10
Rainfall
(Average total)
150
100
50
0
J F M A M J J A S O N D

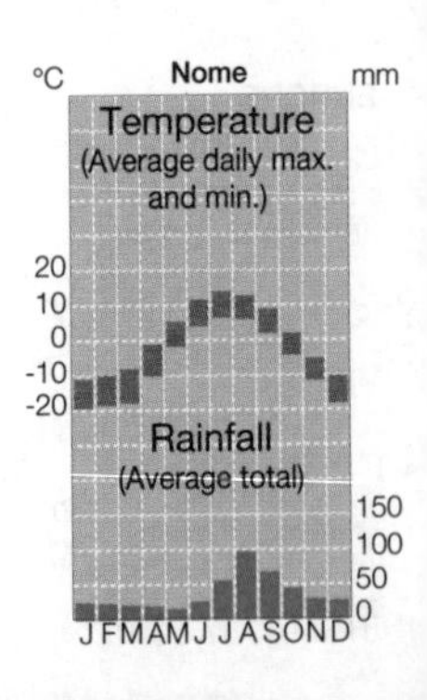
°C
Nome
mm
Temperature
(Average daily max. and min.)
20
10
0
-10
-20
Rainfall
(Average total)
150
100
50
0
J F M A M J J A S O N D

SOUTH AMERICA

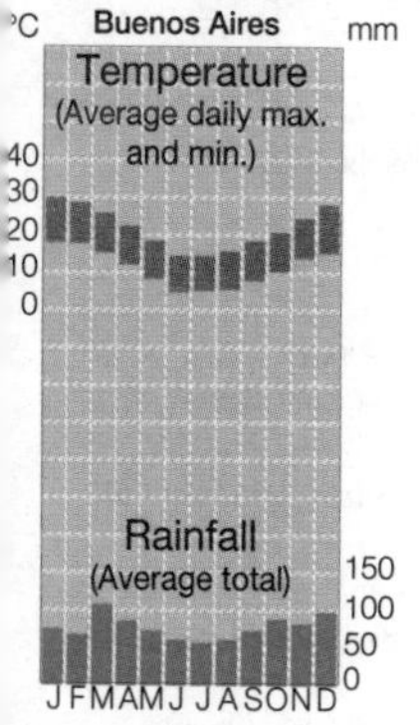

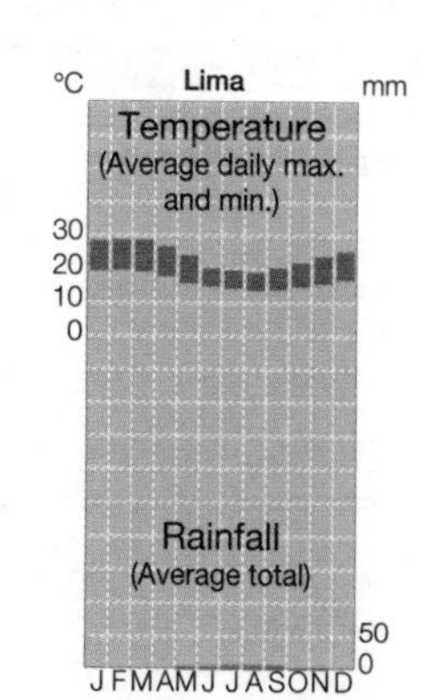

OCEANIA

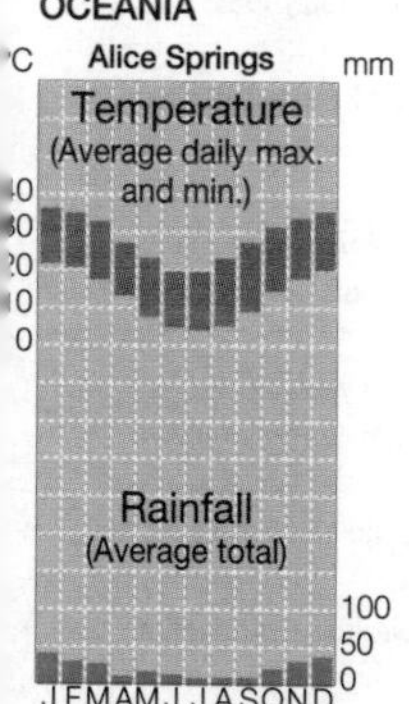

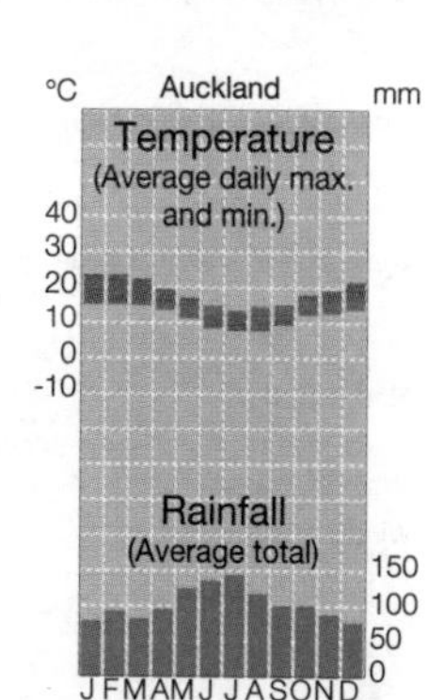

WEATHER EXTREMES

Highest shade temperature	57.8ºC/136ºF Al 'Azīzīyah, Libya (13th September 1922)
Hottest place – Annual mean	34.4ºC/93.9ºF Dalol, Ethiopia
Driest place – Annual mean	0.1 mm/0.004 inches Desierto de Atacama, Chile
Most sunshine – Annual mean	90% Yuma, Arizona, USA (over 4 000 hours)
Least sunshine	Nil for 182 days each year, South Pole
Lowest screen temperature	-89.2ºC/-128.6ºF Vostok Station, Antarctica (21st July 1983)
Coldest place – Annual mean	-56.6ºC/-69.9ºF Plateau Station, Antarctica
Wettest place – Annual mean	11 873 mm/467.4 inches Meghalaya, India
Most rainy days	Up to 350 per year Mount Waialeale, Hawaii, USA
Windiest place	322 km per hour/200 miles per hour in gales, Commonwealth Bay, Antarctica

Highest surface wind speed	
High altitude	372 km per hour/231 miles per hour Mount Washington, New Hampshire, USA (12th April 1934)
Low altitude	333 km per hour/207 miles per hour Qaanaaq (Thule), Greenland (8th March 1972)
Tornado	512 km per hour/318 miles per hour Oklahoma City, Oklahoma, USA (3rd May 1999)
Greatest snowfall	31 102 mm/1 224.5 inches Mount Rainier, Washington, USA (19th February 1971 — 18th February 1972)
Heaviest hailstones	1 kg/2.21 lb Gopalganj, Bangladesh (14th April 1986)
Thunder-days Average	251 days per year Tororo, Uganda
Highest barometric pressure	1 083.8 mb Agata, Siberia, Russian Federation (31st December 1968)
Lowest barometric pressure	870 mb 483 km/300 miles west of Guam, Pacific Ocean (12th October 1979)

TROPICAL STORMS

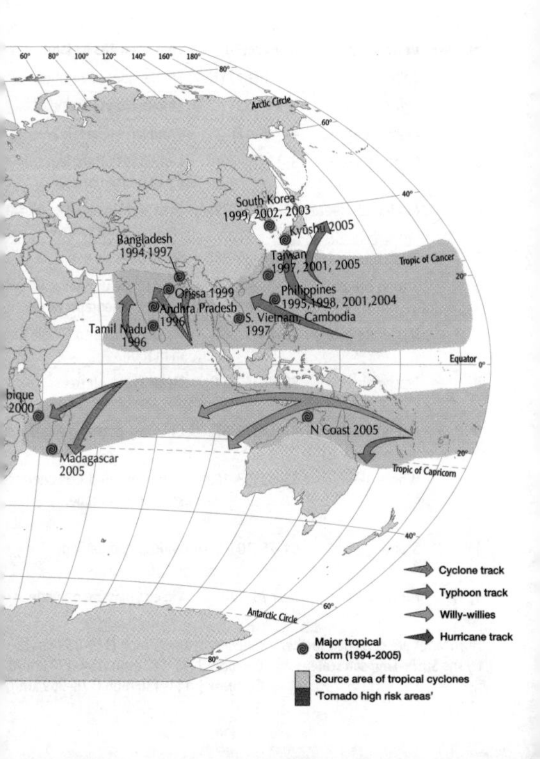

Arctic Circle
South Korea
1999, 2002, 2003
Kyūshū 2005
Bangladesh
1994,1997
Taiwan
1997, 2001, 2005
Tropic of Cancer
Orissa 1999
Philippines
1995,1998, 2001,2004
Andhra Pradesh
1996
S. Vietnam, Cambodia
1997
Tamil Nadu
1996
Equator
bique
2000
N Coast 2005
Madagascar
2005
Tropic of Capricorn
Antarctic Circle
Cyclone track
Typhoon track
Willy-willies
Hurricane track
Major tropical
storm (1994-2005)
Source area of tropical cyclones
'Tornado high risk areas'

BEAUFORT SCALE OF WIND SPEED

Scale number	Description	Speed mph (km/h)	Characteristics on land
0	Calm	Less than 1 (1)	Smoke goes straight up.
1	Light air	1–3 (1–5)	Smoke blows in wind.
2	Light breeze	4–7 (6–12)	Wind felt on face; leaves rust
3	Gentle breeze	8–12 (13–20)	Light flag flutters; leaves in constant motion.
4	Moderate breeze	13–18 (21–29)	Dust and loose paper blown. Small branches move.
5	Fresh breeze	19–24 (30–39)	Small trees sway.
6	Strong breeze	25–31 (40–50)	Hard to use umbrellas. Whistl heard in telegraph wires.
7	Moderate gale	32–38 (51–61)	Hard to walk into. Whole tree in motion.
8	Fresh gale	39–46 (62–74)	Twigs break off trees.
9	Strong gale	47–54 (75–87)	Chimney pots and slates lost
10	Whole gale	55–63 (88–102)	Trees uprooted. Considerabl structural damage.
11	Storm	64–75 (103–120)	Widespread damage.
12–17	Hurricane	over 75 (120)	Violent, massive damage.

Hurricanes are classified further by the **Saffir-Simpson scale:**

Category 1 74–95 mph (119–153 km/h)
Category 2 96–110 mph (154–177 km/h)
Category 3 111–130 mph (178–209 km/h

Characteristics at sea
Sea like a mirror.
Ripples formed, but without foam crests.
Small wavelets. Crests glassy but do not break.
Large wavelets. Crests begin to break. Foam glassy, scattered white horses.
Small waves. Fairly frequent white horses.
Moderate waves. Many white horses. Chance of spray.
Large waves, extensive white foam crests. Probably spray.
Sea leaps up. White foam from breaking waves begins to be blown in streaks along wind direction.
Moderately high waves. Edges of crests begin to break into the spindrift. Foam blown in well-marked streaks along wind direction.
High waves. Dense streaks of foam along wind direction. Crests topple, tumble and roll over. Spray may affect visibility.
Very high waves with long overhanging crests. Foam is blown in great patches along wind direction. Surface takes on a general white appearance. Visibility affected.
Exceptionally high waves. Sea completely covered with long white patches of foam lying along wind direction. Visibility affected.
Air is filled with foam and spray. Sea completely white with driving spray. Visibility seriously affected.
Category 4 131–155 mph (210–249 km/h) *Category 5* over 155 mph (over 249 km/h)

POPULATION AND CITIES

POPULATION

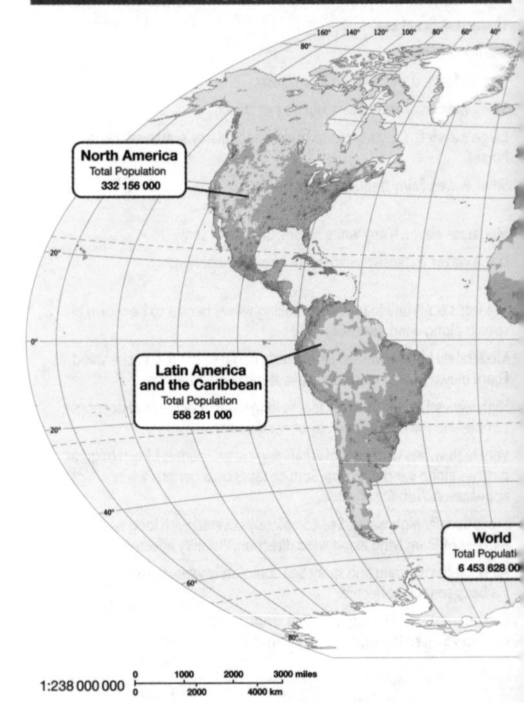

WORLD POPULATION DISTRIBUTION

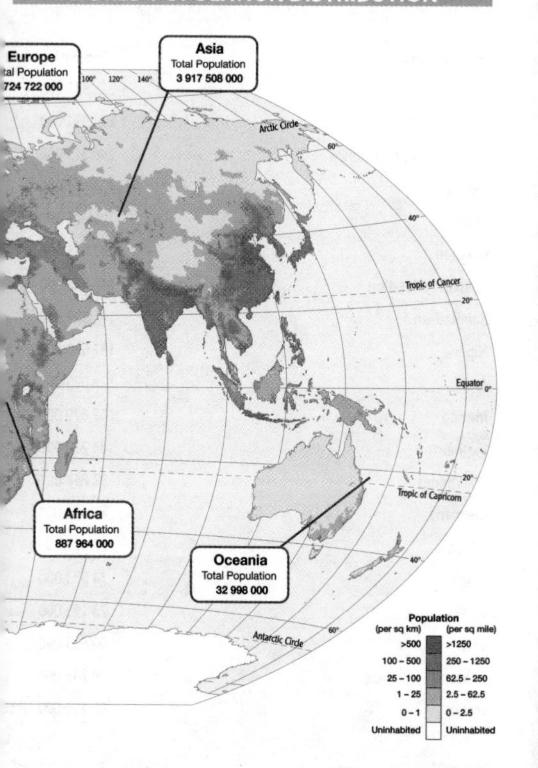
Europe
tal Population
724 722 000
Asia
Total Population
3 917 508 000
100°
120°
140°
Arctic Circle
60°
40°
Tropic of Cancer
20°
Equator 0°
20°
Tropic of Capricorn
40°
Africa
Total Population
887 964 000
Oceania
Total Population
32 998 000
60°
Antarctic Circle
Population
(per sq km)
(per sq mile)
>500
>1250
100 – 500
250 – 1250
25 – 100
62.5 – 250
1 – 25
2.5 – 62.5
0 – 1
0 – 2.5
Uninhabited
Uninhabited

WORLD: POPULATION

HIGHEST POPULATIONS	Population
China	1 323 345 000
India	1 103 371 000
United States of America	298 213 000
Indonesia	222 781 000
Brazil	186 405 000
Pakistan	157 935 000
Russian Federation	143 202 000
Bangladesh	141 822 000
Nigeria	131 530 000
Japan	128 085 000
Mexico	107 029 000
Vietnam	84 238 000
Philippines	83 054 000
Germany	82 689 000
Ethiopia	77 431 000
Egypt	74 033 000
Turkey	73 193 000
Iran	69 515 000
Thailand	64 233 000
France	60 496 000

LOWEST POPULATIONS	Population
St Lucia	161 000
São Tomé and Príncipe	157 000
St Vincent and the Grenadines	119 000
Federated States of Micronesia	110 000
Grenada	103 000
Tonga	102 000
Kiribati	99 000
Antigua and Barbuda	81 000
Seychelles	81 000
Dominica	79 000
Andorra	67 000
Marshall Islands	62 000
St Kitts and Nevis	43 000
Liechtenstein	35 000
Monaco	35 000
San Marino	28 000
Palau	20 000
Nauru	14 000
Tuvalu	10 000
Vatican City	552

WORLD: POPULATION DENSITIES

HIGHEST POPULATION DENSITIES	People per sq km	People per sq mile
Monaco	17 500.0	35 000.0
Singapore	6 770.0	17 514.2
Malta	1 272.2	3 295.1
Maldives	1 104.0	2 860.9
Vatican City	1 104.0	2 860.9
Bahrain	1 052.1	2 722.8
Bangladesh	984.9	2 550.8
Nauru	666.7	1 750.0
Taiwan	631.8	1 636.3
Barbados	627.9	1 626.5
Mauritius	610.3	1 579.9
South Korea	481.7	1 247.5
San Marino	459.0	1 166.7
Comoros	428.6	1 109.9
Tuvalu	400.0	1 000.0
Netherlands	392.5	1 016.6
India	360.0	932.4
Rwanda	343.2	888.8
Marshall Islands	342.5	885.7
Lebanon	342.2	886.3

LOWEST POPULATION DENSITIES	People per sq km	People per sq mile
Niger	11.0	28.5
Mali	10.9	28.2
Turkmenistan	9.9	25.6
Russian Federation	8.4	21.7
Bolivia	8.4	21.7
Oman	8.3	21.5
Chad	7.6	19.7
Central African Republic	6.5	16.8
Kazakhstan	5.5	14.1
Gabon	5.2	13.4
Guyana	3.5	9.0
Libya	3.3	8.6
Canada	3.2	8.4
Botswana	3.0	7.9
Mauritania	3.0	7.9
Iceland	2.9	7.4
Suriname	2.7	7.1
Australia	2.6	6.8
Namibia	2.5	6.4
Mongolia	1.7	4.4

KEY POPULATION STATISTICS

	Population 2005 (millions)	Growth (per cent)	Infant mortality rate
World	6 454	1.2	57
More developed regions[1]	1 209	0.3	8
Less developed regions[2]	5 244	1.4	62
Africa	888	2.2	94
Asia	3 918	1.2	54
Europe[3]	725	0.0	9
Latin America and the Caribbean[4]	558	1.4	26
North America	332	1.0	7
Oceania	33	1.3	29

[1] Europe, North America, Australia, New Zealand and Japan.

[2] Africa, Asia (excluding Japan), Latin America and the Caribbean, and Oceani

[3] Includes Russian Federation.

[4] South America, Central America (including Mexico) and all Caribbean Island

Total fertility rate	Life expectancy (years)	% aged 60 and over	
		2005	2050
2.7	65	10	22
1.6	76	20	32
2.9	63	8	20
5.0	49	5	10
2.5	67	9	24
1.4	74	21	34
2.6	72	9	24
2.0	78	17	27
2.3	74	14	25

ding Australia and New Zealand).

CITIES

DISTRIBUTION OF THE WORLD'S MAJOR CITIES

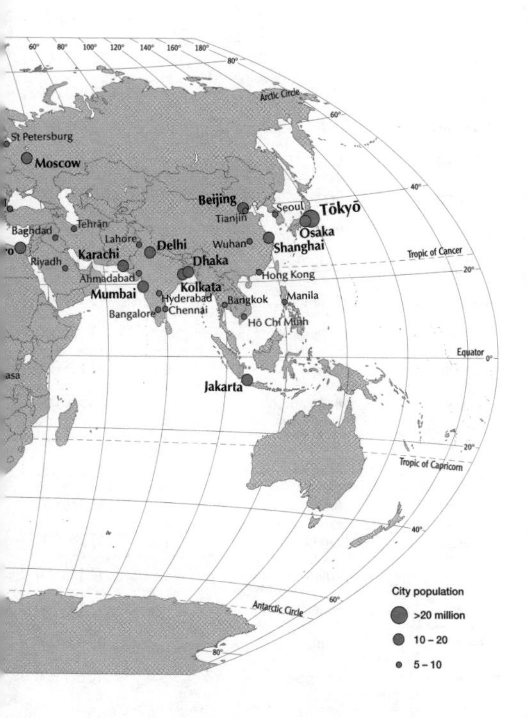
St Petersburg
Moscow
Beijing
Tianjin
Seoul
Tōkyō
Osaka
Tehran
Baghdad
Lahore
Delhi
Wuhan
Shanghai
Karachi
Riyadh
Dhaka
Ahmadabad
Hong Kong
Mumbai
Kolkata
Hyderabad
Bangkok
Manila
Bangalore
Chennai
Hô Chi Minh
Jakarta
Arctic Circle
Tropic of Cancer
Equator
Tropic of Capricorn
Antarctic Circle
City population
>20 million
10 – 20
5 – 10

THE WORLD'S LARGEST CITIES

		Population
Tōkyō	Japan	35 327 000
México City	Mexico	19 013 000
New York	USA	18 498 000
Mumbai	India	18 336 000
São Paulo	Brazil	18 333 000
Delhi	India	15 334 000
Kolkata	India	14 299 000
Buenos Aires	Argentina	13 349 000
Jakarta	Indonesia	13 194 000
Shanghai	China	12 665 000
Dhaka	Bangladesh	12 560 000
Los Angeles	USA	12 146 000
Karachi	Pakistan	11 819 000
Rio de Janeiro	Brazil	11 469 000
Ōsaka	Japan	11 286 000
Cairo	Egypt	11 146 000
Lagos	Nigeria	11 135 000
Beijing	China	10 849 000
Manila	Philippines	10 677 000

		Population
Moscow	Russian Federation	10 672 000
Paris	France	9 854 000
İstanbul	Turkey	9 760 000
Seoul	South Korea	9 592 000
Tianjin	China	9 346 000
Chicago	USA	8 711 000
Lima	Peru	8 180 000
London	United Kingdom	7 615 000
Bogotá	Colombia	7 594 000
Tehrān	Iran	7 352 000
Hong Kong	China	7 182 000
Chennai	India	6 915 000
Bangkok	Thailand	6 604 000
Essen	Germany	6 566 000
Bangalore	India	6 532 000
Lahore	Pakistan	6 373 000
Hyderabad	India	6 145 000
Wuhan	China	6 003 000
Baghdād	Iraq	5 910 000
Kinshasa	Democratic Republic of the Congo	5 717 000
Santiago	Chile	5 623 000

GROWTH OF CITIES OF OVER 10 MILLION

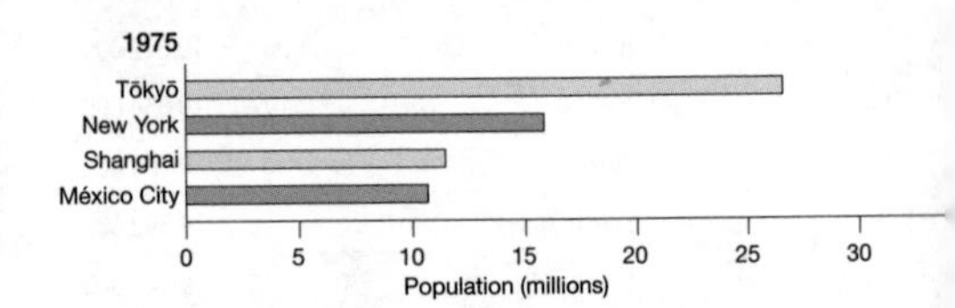

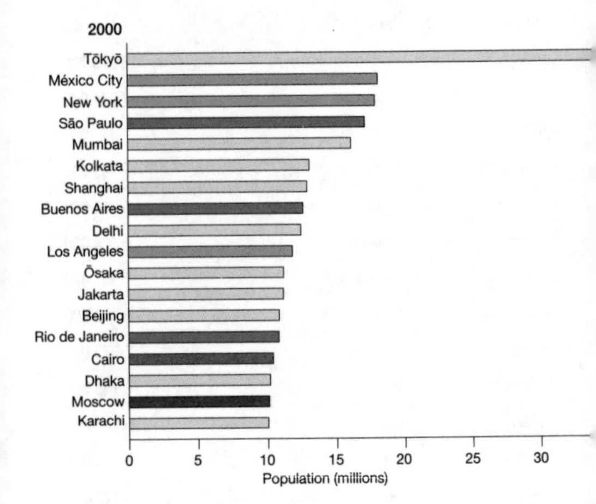

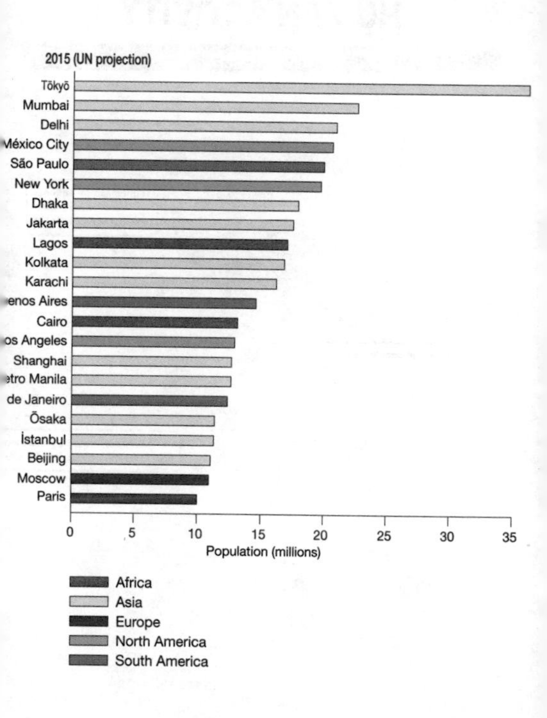
2015 (UN projection)
Tōkyō
Mumbai
Delhi
México City
São Paulo
New York
Dhaka
Jakarta
Lagos
Kolkata
Karachi
enos Aires
Cairo
os Angeles
Shanghai
etro Manila
de Janeiro
Ōsaka
İstanbul
Beijing
Moscow
Paris
0
5
10
15
20
25
30
35
Population (millions)
Africa
Asia
Europe
North America
South America

HUMAN ACTIVITY

WEALTH

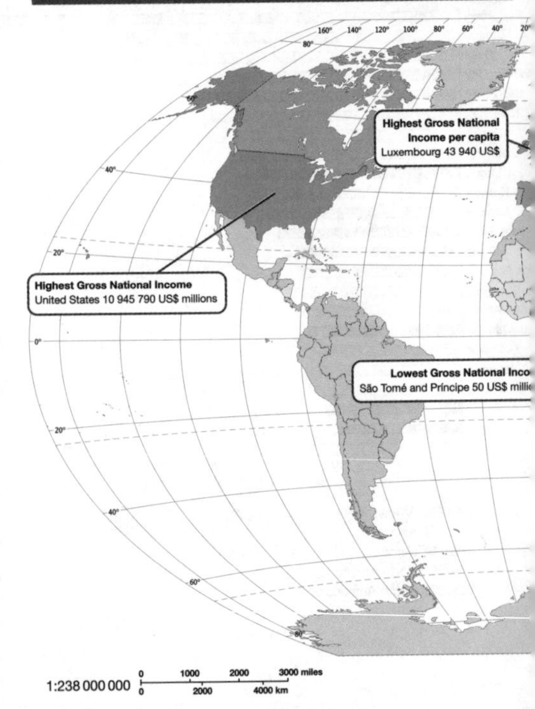
Highest Gross National
Income per capita
Luxembourg 43 940 US$
Highest Gross National Income
United States 10 945 790 US$ millions
Lowest Gross National Inco
São Tomé and Príncipe 50 US$ millio
1:238 000 000
0 1000 2000 3000 miles
0 2000 4000 km

GROSS NATIONAL INCOME

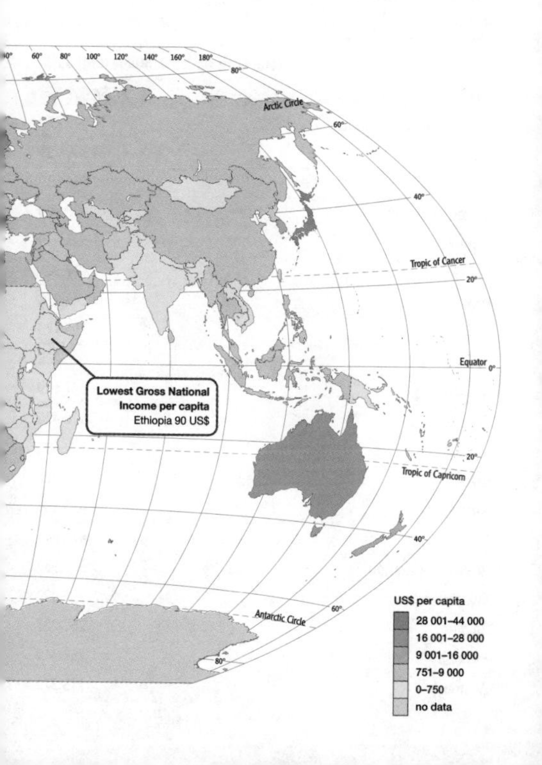

RICHEST AND POOREST COUNTRIES

HIGHEST TOTAL WEALTH

	Total Gross National Income (GNI) US$M
World	34 491 458
United States of America	10 945 790
Japan	4 389 791
Germany	2 084 631
United Kingdom	1 680 300
France	1 523 025
China	1 417 301
Italy	1 242 978
Canada	756 770
Spain	698 208
Mexico	637 159
South Korea	576 426
India	567 605
Brazil	478 922
Australia	430 533
Netherlands	426 641
Russian Federation	374 937
Switzerland	292 892
Belgium	267 227
Sweden	258 319
Austria	215 372

LOWEST TOTAL WEALTH

	Total Gross National Income (GNI) US$M
Bhutan	578
Liberia	445
The Gambia	442
Equatorial Guinea	437
Grenada	396
St Vincent and the Grenadines	361
East Timor	351
St Kitts and Nevis	321
Samoa	284
Solomon Islands	273
Comoros	269
Federated States of Micronesia	261
Vanuatu	248
Dominica	239
Guinea-Bissau	202
Tonga	152
Palau	150
Marshall Islands	143
Kiribati	84
São Tomé and Príncipe	50

RICHEST AND POOREST COUNTRIES

RICHEST COUNTRIES

	Gross National Income (GNI) per capita (US$)
Luxembourg	43 940
Norway	43 350
Switzerland	39 880
United States of America	37 610
Japan	34 510
Denmark	33 750
Iceland	30 810
Sweden	28 840
United Kingdom	28 350
Finland	27 020
Ireland	26 960
Austria	26 720
Netherlands	26 310
Belgium	25 820
Germany	25 250
France	24 770
Canada	23 930
Australia	21 650
Italy	21 560
Singapore	21 230

POOREST COUNTRIES

	Gross National Income (GNI) per capita (US$)
Burkina	300
Madagascar	290
Mali	290
Tanzania	290
Central African Republic	260
Chad	250
Nepal	240
Uganda	240
Rwanda	220
Mozambique	210
Niger	200
Eritrea	190
Tajikistan	190
Malawi	170
Sierra Leone	150
Guinea-Bissau	140
Liberia	130
Burundi	100
Democratic Republic of the Congo	100
Ethiopia	90

INTERNATIONAL AID

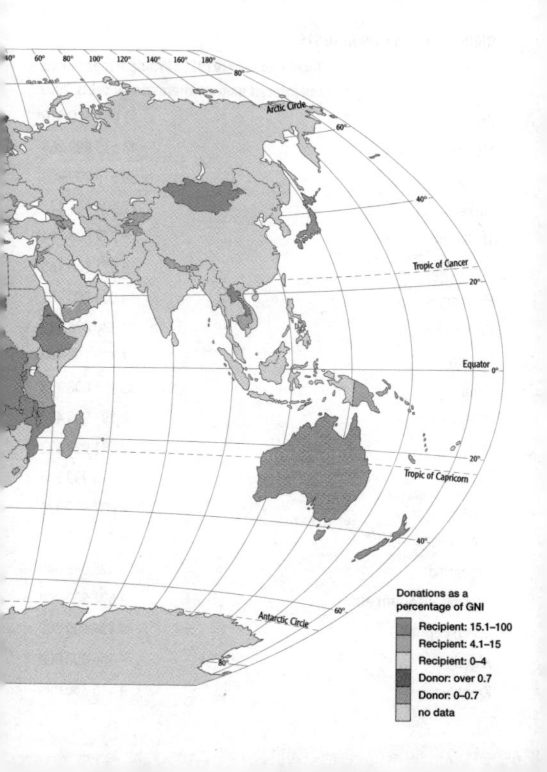
60°
80°
100°
120°
140°
160°
180°
80°
Arctic Circle
60°
40°
Tropic of Cancer
20°
Equator 0°
20°
Tropic of Capricorn
40°
60°
Antarctic Circle
80°
Donations as a percentage of GNI
Recipient: 15.1–100
Recipient: 4.1–15
Recipient: 0–4
Donor: over 0.7
Donor: 0–0.7
no data

INTERNATIONAL DEBT

HIGHEST INTERNATIONAL DEBTS

	Total debt service (US$): Principal repayments and interest paid on international debts, 2002
Brazil	51 631 599 616
Mexico	43 535 499 264
China	30 615 799 808
Turkey	27 604 400 128
Thailand	19 737 800 704
Indonesia	16 971 100 160
Hungary	14 869 900 288
Russian Federation	14 330 099 712
Poland	13 488 799 744
India	13 127 600 128
Philippines	9 191 999 488
Malaysia	8 081 999 872
Chile	7 728 999 936
Venezuela	7 487 399 936
Colombia	6 920 699 904
Argentina	5 825 699 840
Republic of South Africa	4 691 500 032
Czech Republic	4 534 400 000
Algeria	4 166 400 000
Kazakhstan	4 115 300 096

HIGHEST DEBT REPAYMENTS

	Debt Service Ratio: Percentage of Gross National Income (GNI) used to pay off international debts, 2002
Hungary	24.3
Belize	22.7
Kazakhstan	17.4
Democratic Republic of the Congo	16.8
Thailand	15.8
Turkey	15.2
Slovakia	14.2
Panama	13.9
Croatia	13.6
São Tomé and Príncipe	13.1
Estonia	12.7
Moldova	12.6
St Kitts and Nevis	12.4
Lebanon	12.2
Chile	11.9
Brazil	11.7
Guyana	11.6
Jamaica	11.6
Kyrgyzstan	11.2
Philippines	11.1

SOCIAL INDICATORS

INFANT MORTALITY

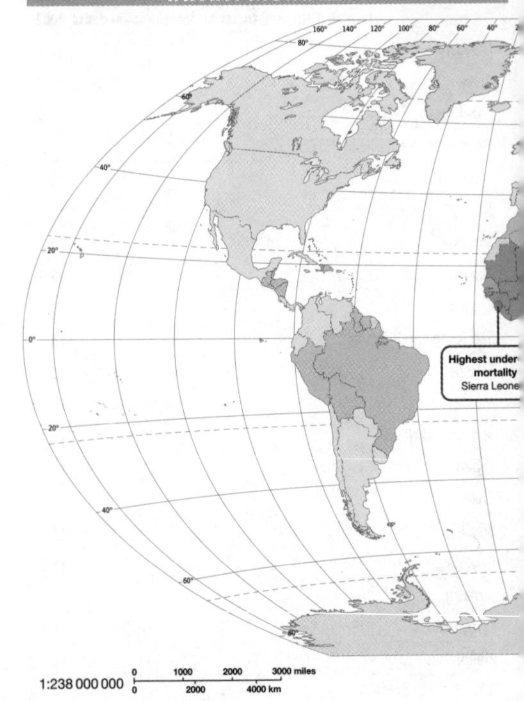

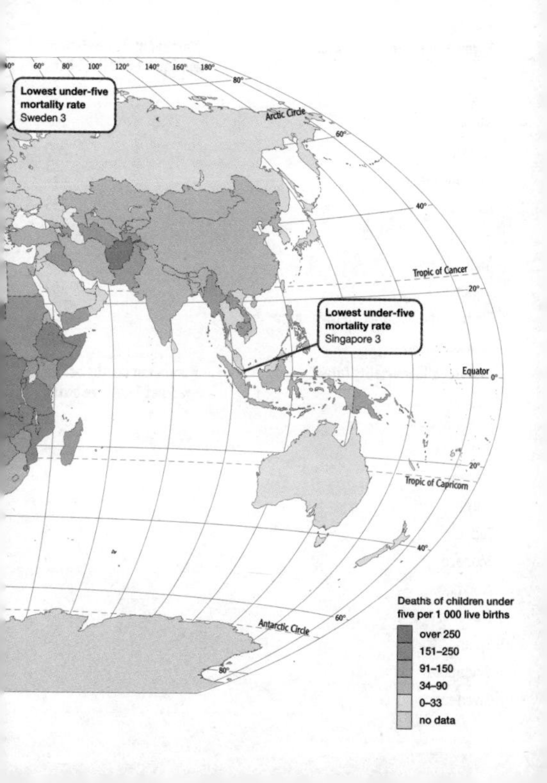
Lowest under-five mortality rate
Sweden 3
Lowest under-five mortality rate
Singapore 3
Arctic Circle
Tropic of Cancer
Equator
Tropic of Capricorn
Antarctic Circle
Deaths of children under five per 1 000 live births
over 250
151–250
91–150
34–90
0–33
no data

INFANT MORTALITY

Highest child mortality rates	Number of deaths before age 5 per 1 000 live births
Sierra Leone	284
Niger	262
Angola	260
Afghanistan	257
Liberia	235
Somalia	225
Mali	220
Burkina	207
Democratic Republic of the Congo	205
Guinea-Bissau	204
Lowest child mortality rates	**Number of deaths before age 5 per 1 000 live births**
Czech Republic	4
Denmark	4
Iceland	4
Italy	4
Japan	4
Monaco	4
Norway	4
Slovenia	4
Spain	4
Singapore	3
Sweden	3

CAUSES OF DEATH

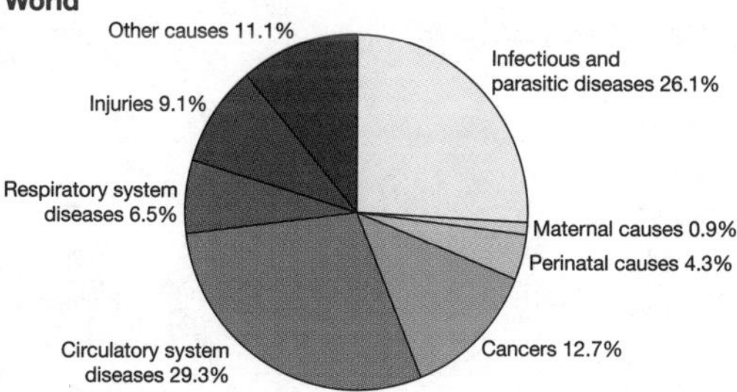

Total : 56 937 349

Developing countries

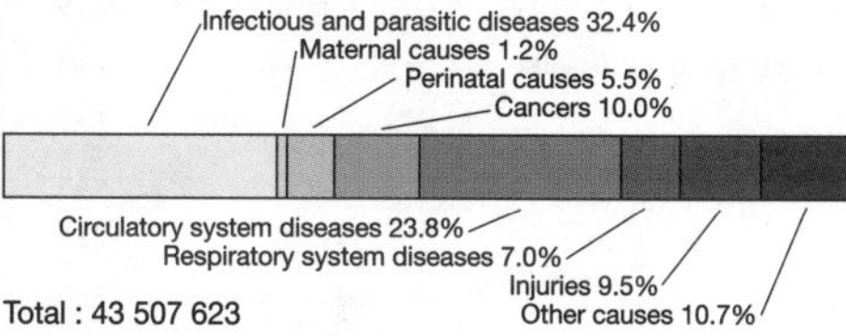

Total : 43 507 623

Developed countries

Infectious and parasitic diseases 5.5%
Maternal causes 0.0%
Perinatal causes 0.6%
Cancers 21.6%

Other causes 12.3%
Injuries 7.8%
Respiratory system diseases 4.9%
Circulatory system diseases 12.3%

Total : 13 429 726

LIFE EXPECTANCY

Highest life expectancy (male)	years
Japan	77.9
Iceland	77.6
Sweden	77.6
Israel	77.1
Canada	76.7
Australia	76.4
Cyprus	76.0
Norway	76.0
Malta	75.9
Switzerland	75.9
Highest life expectancy (female)	**years**
Japan	85.1
France	82.8
Spain	82.8
Sweden	82.6
Switzerland	82.3
Australia	82.0
Belgium	81.9
Canada	81.9
Iceland	81.9
Norway	81.9

Lowest life expectancy (male)	years
Angola	38.8
Rwanda	38.8
Central African Republic	38.5
Malawi	37.3
Mozambique	36.6
Zimbabwe	33.7
Swaziland	33.3
Sierra Leone	33.1
Zambia	32.7
Lesotho	32.3
Lowest life expectancy (female)	**years**
Central African Republic	40.6
Botswana	40.5
Rwanda	39.7
Mozambique	39.6
Lesotho	37.7
Malawi	37.7
Sierra Leone	35.5
Swaziland	35.4
Zimbabwe	32.6
Zambia	32.1

HEALTH PROVISION

AVAILABILITY OF DOCTORS

Best provision	Number of doctors per 100 000 people
Italy	607
Cuba	596
Georgia	463
Belarus	450
Greece	438
Russian Federation	420
Belgium	419
Lithuania	403
Uruguay	387
Israel	375
Worst provision	**Number of doctors per 100 000 people**
Burkina	4
Central African Republic	4
Mali	4
Tanzania	4
Chad	3
Ethiopia	3
Niger	3
Mozambique	2
Rwanda	2
Burundi	1

HEALTH SPENDING	
Highest spenders	**Health spending as percentage of Gross Domestic Product**
USA	13.9
Lebanon	12.2
Cambodia	11.8
Switzerland	11
Uruguay	10.9
Germany	10.8
East Timor	9.8
Marshall Islands	9.8
France	9.6
Jordan	9.5
Lowest spenders	**Health spending as percentage of Gross Domestic Product**
Chad	2.6
Somalia	2.6
North Korea	2.5
Indonesia	2.4
São Tomé and Príncipe	2.3
Congo	2.1
Myanmar	2.1
Equatorial Guinea	2.0
Madagascar	2.0
Azerbaijan	1.6

SAFE WATER

Lowest access to safe water	Percentage of population with access to safe drinking water
East Timor	52
Swaziland	52
Burkina	51
Guinea	51
Togo	51
Angola	50
Mali	48
Congo	46
Democratic Republic of the Congo	46
Niger	46
Madagascar	45
Equatorial Guinea	44
Laos	43
Mozambique	42
Papua New Guinea	39
Cambodia	34
Chad	34
Somalia	29
Ethiopia	22
Afghanistan	13

UN MILLENNIUM DEVELOPMENT GOALS

From the Millennium Declaration, 2000

Goal 1	Eradicate extreme poverty and hunger
Goal 2	Achieve universal primary education
Goal 3	Promote gender equality and empower women
Goal 4	Reduce child mortality
Goal 5	Improve maternal health
Goal 6	Combat HIV/AIDS, malaria and other diseases
Goal 7	Ensure environmental sustainability
Goal 8	Develop a global partnership for development

The UN Millennium Development Goals were issued in September 2000. Forty-eight specific indicators were identified to measure the progress of countries trying to combat the development issues being addressed by the eight goals. These issues include those of poverty, hunger, health, water supply and environmental degradation.

ENVIRONMENT

WORLD: LARGEST PROTECTED AREAS

	Location
Greenland	Greenland
Rub` al Khālī	Saudi Arabia
Great Barrier Reef Marine Park	Australia
Northwestern Hawaiian Islands	United States
Amazonia	Colombia
Qiangtang	China
Macquarie Island	Australia
Sanjiangyuan	China
Cape Churchill	Canada
Galapagos Islands	Ecuador
Northern Wildlife Management Zone	Saudi Arabia
Ngaanyatjarra Lands	Australia
Alto Orinoco-Casiquiare	Venezuela
Vale do Javari	Brazil
Ouadi Rimé-Ouadi Achim	Chad
Arctic	United States
Yanomami	Brazil
Yukon Delta	United States
Aïr and Ténéré	Niger
Pacifico	Colombia

Area (sq km)	Area (sq miles)	Designation
972 000	375 289	National Park
640 000	247 104	Wildlife Management Area
344 360	132 957	Marine Park
341 362	131 800	Coral Reef Ecosystem Reserve
326 329	125 996	Forest Reserve
298 000	115 058	Nature Reserve
162 060	62 571	Marine Park
152 300	58 803	Nature Reserve
137 072	52 923	Wildlife Management Area
133 000	51 351	Marine Reserve
100 875	38 948	Wildlife Management Area
98 129	37 888	Indigeneous Protected Area
84 000	32 432	Biosphere Reserve
83 380	32 193	Indigenous Area
80 000	30 888	Faunal Reserve
78 049	30 135	National Wildlife Refuge
77 519	29 930	Indigenous Park
77 425	29 894	National Wildlife Refuge
77 360	29 869	National Nature Reserve
73 981	28 564	Forest Reserve

WORLD HERITAGE

The UNESCO World Heritage Convention

In 1972, the United Nations Educational, Scientific and Cultural Organization (UNESCO) adopted an international treaty called the 'Convention concerning the Protection of the World Cultural and Natural Heritage'. This aims to assist nations in identifying, protecting, preserving and managing cultural and natural heritage sites worldwide. Such places are unique, irreplaceable sites demonstrating either man's achievement or nature's creation. The Convention defines the kind of sites which can be considered for inclusion on the World Heritage List.

The World Heritage Committee meets once a year to decide which sites should be added to this list. There are currently 812 World Heritage Sites located in 137 countries around the world. Of these, 628 are classified as cultural, 160 natural and 24 mixed.

Selection criteria

To be included on the list, sites must be of outstanding universal value and meet at least one out of ten selection criteria:

i. to represent a masterpiece of human creative genius;

ii. to exhibit an important interchange of human values, over a span of time or within a cultural area of the world, on developments in architecture or technology, monumental arts, town-planning or landscape design;

iii. to bear a unique or at least exceptional testimony to a cultural tradition or to a civilization which is living or which has disappeared;

iv. to be an outstanding example of a type of building, architectural or technological ensemble or landscape which illustrates (a) significant stage(s) in human history;

v. to be an outstanding example of a traditional human settlement, land-use, or sea-use which is representative of a culture (or cultures), or human interaction with the environment especially when it has become vulnerable under the impact of irreversible change;

vi. to be directly or tangibly associated with events or living traditions, with ideas, or with beliefs, with artistic and literary works of outstanding universal significance. (The Committee considers that this criterion should preferably be used in conjunction with other criteria);

vii. to contain superlative natural phenomena or areas of exceptional natural beauty and aesthetic importance;

viii. to be outstanding examples representing major stages of earth's history, including the record of life, significant on-going geological processes in the development of landforms, or significant geomorphic or physiographic features;

ix. to be outstanding examples representing significant on-going ecological and biological processes in the evolution and development of terrestrial, fresh water, coastal and marine ecosystems and communities of plants and animals;

x. to contain the most important and significant natural habitats for in-situ conservation of biological diversity, including those containing threatened species of outstanding universal value from the point of view of science or conservation.

NATURAL WORLD HERITAGE SITES

Natural sites
Mixed sites

CONFLICT

REFUGEES AND MILITARY SPENDING

MAJOR REFUGEE POPULATIONS

Country of Origin [1]	Main Countries of Asylum	Total
Afghanistan	Pakistan / Iran	2 136 000
Sudan	Uganda / Chad / Ethiopia / Kenya / Democratic Republic of the Congo / Central African Republic	606 200
Burundi	Tanzania / Democratic Republic of the Congo / Zambia / South Africa / Rwanda	531 600
Democratic Republic of the Congo	Tanzania / Congo / Zambia / Burundi / Rwanda / Angola / Uganda	453 400
Palestinians [2]	Saudi Arabia / Iraq / Egypt / Libya / Algeria	427 900
Somalia	Kenya / Yemen / United Kingdom / Ethiopia / Djibouti / USA	402 200
Iraq	Iran / Germany / Netherlands / Sweden / United Kingdom	368 500
Vietnam	China / Germany / USA / France	363 200
Liberia	Guinea / Côte d'Ivoire / Sierra Leone / Ghana / USA	353 300
Angola	Zambia / Democratic Republic of the Congo / Namibia / South Africa	329 600

[1] This table includes UNHCR estimates for nationalities in industrialized countries on the basis of recent refugee arrivals and asylum seeker recognition.

[2] This figure excludes some 4 million Palestinians who are covered by a separate mandate of the U.N. Relief and Works Agency for Palestine Refugees in the Near East (UNRWA).

MILITARY SPENDING

Highest spenders	Percentage of Gross Domestic Product spent on military activities
Oman	13.0
Saudi Arabia	11.3
Kuwait	11.2
Bosnia-Herzegovina	9.5
Israel	8.6
Jordan	8.4
Burundi	7.6
Brunei	7.0
Syria	6.1
Ethiopia	5.2
Lowest spenders	**Percentage of Gross Domestic Product spent on military activities**
Cape Verde	0.7
Ireland	0.7
Georgia	0.6
Ghana	0.6
Guatemala	0.6
Zambia	0.6
Mexico	0.5
Moldova	0.3
Mauritius	0.2
Iceland	0.0

TERRORISM

Major terrorist incidents	Date
Lockerbie, Scotland	December 1988
Tōkyō, Japan	March 1995
Oklahoma City, USA	April 1995
Nairobi, Kenya and Dar es Salaam, Tanzania	August 1998
Omagh, Northern Ireland	August 1998
New York and Washington D.C., USA	September 2001
Bali, Indonesia	October 2002
Moscow, Russian Federation	October 2002
Bāghdad and Karbalā', Iraq	March 2004
Madrid, Spain	March 2004
Beslan, Russian Federation	September 2004
London, UK	July 2005
Sharm ash Shaykh, Egypt	July 2005

Terrorist Incidents by Region 1998–2005

	Incidents	Injuries	Fatalities
Middle East	4 355	12 814	5 385
Western Europe	2 672	1 463	336
South Asia	2 541	10 404	4 173
Latin America	1 461	2 095	1 341
Eastern Europe	1 035	4 664	1 808
Southeast Asia and Oceania	431	3 138	918
Africa	311	7 249	2 127
North America	103	30	2 994
East and Central Asia	83	151	134

Summary	Killed	Injured
Airline bombing	270	5
Sarin gas attack on subway	12	5 700
Bomb in the Federal building	168	over 500
US Embassy bombings	257	over 4 000
Town centre bombing	29	330
Airline hijacking and crashing	2 752	4 300
Car bomb outside nightclub	202	300
Theatre siege	170	over 600
Suicide bombing of pilgrims	181	over 400
Train bombings	191	1 800
School siege	330	700
Underground and bus bombings	52	700
Bombs at tourist sites	88	200

Web links

U.K. Foreign and Commonwealth Office (FCO)	www.fco.gov.uk
U.S. Department of State, Counterterrorism Office	www.state.gov/s/ct
National Memorial Institute for the Prevention of Terrorism (MIPT)	www.mipt.org
National Counterterrorism Center (NCTC)	www.nctc.gov
Terrorism Research Center, Inc. (TRC)	www.terrorism.com

TRAVEL

WORLD'S BUSIEST AIRPORTS

	Location	Number of passengers 1977	Number of passengers 2004
Atlanta Hartsfield-Jackson	USA	29 977 465	83 606 583
Chicago O'Hare	USA	n/a	75 533 822
London Heathrow	UK	23 775 605	67 344 054
Tokyo Haneda	Japan	23 641 455	62 291 405
Los Angeles	USA	28 361 863	60 688 609
Dallas/Fort Worth	USA	17 318 728	59 412 217
Paris Charles de Gaulle	France	n/a	51 260 363
Frankfurt	Germany	14 968 337	51 098 271
Amsterdam Schiphol	Netherlands	8 931 985	42 541 180
Denver	USA	15 281 842	42 393 766
Las Vegas McCarran	USA	n/a	41 441 531
Phoenix Sky Harbor	USA	5 931 806	39 504 898
Madrid Barajas	Spain	9 378 344	38 704 731
Bangkok	Thailand	3 922 569	37 960 169

	Location	Number of passengers 1977	Number of passengers 2004
New York JFK	USA	22 545 497	37 518 143
Minneapolis/St Paul	USA	n/a	36 713 173
Hong Kong	China	5 444 821	36 711 920
Houston George Bush Intercontinental	USA	7 996 935	36 506 116
Detroit Metropolitan Wayne County	USA	4 361 523	35 187 517
Beijing	China	n/a	34 883 190
San Francisco	USA	19 498 146	32 247 746
New York Newark	USA	n/a	31 947 266
London Gatwick	UK	6 652 336	31 461 454
Orlando	USA	4 154 781	31 143 388
Tokyo Narita	Japan	7 262 000	31 057 252
Singapore Changi	Singapore	n/a	30 353 565
Miami	USA	13 736 483	30 165 197
Seattle-Tacoma	USA	n/a	28 804 554
Toronto Lester B. Pearson	Canada	7 056 949	28 615 709
Philadelphia	USA	n/a	28 507 420

WORLD'S BUSIEST AIR ROUTES

	Annual number of passengers (both directions)
London–New York	3 832 630
London–Amsterdam	3 334 541
London–Paris	2 655 116
Hong Kong–T'aipei	2 617 471
Seoul–Tōkyō	2 314 950
Kuala Lumpur–Singapore	2 166 985
Hong Kong–Bangkok	2 158 923
London–Dublin	2 019 221
Bangkok–Singapore	2 014 392
London–Frankfurt	1 943 328
Hong Kong–Tōkyō	1 813 558
Honolulu–Tōkyō	1 796 052
Hong Kong–Singapore	1 592 101
Hong Kong–Manila	1 557 552
New York–Paris	1 537 225
Jakarta–Singapore	1 499 975
London–Madrid	1 408 868
Madrid–Paris	1 378 871
Ōsaka–Seoul	1 362 012
New York–Toronto	1 342 454

	Annual number of passengers (both directions)
London–Los Angeles	1 318 627
Tōkyō–Bangkok	1 317 245
Hong Kong–Seoul	1 307 414
London–Chicago	1 270 937
Singapore–Tōkyō	1 230 769
London–Brussels	1 193 121
Frankfurt–New York	1 182 741
Tōkyō–Los Angeles	1 114 101
London–Boston	1 109 692
London–Barcelona	1 105 671
London–Zurich	1 100 064
Frankfurt–Paris	1 078 592
London–Stockholm	1 072 624
London–Munich	1 064 205
Tōkyō–T'aipei	1 058 265
Chicago–Toronto	1 033 936
London–San Francisco	1 012 837
London–Nice	971 248
London–Washington	968 510
Paris–Amsterdam	966 565

DISTANCE CHART

Beijing					
19 300 *11 993*	**Buenos Aires**				
5 810 *3 610*	13 600 *8 451*	**Dubai**			
1 960 *1 218*	18 500 *11 496*	5 940 *3 691*	**Hong Kong**		
11 700 *7 270*	8 110 *5 040*	6 410 *3 983*	10 700 *6 649*	**Johannesburg**	
8 140 *5 058*	11 100 *6 898*	5 490 *3 411*	9 640 *5 990*	9 060 *5 630*	**London**
10 100 *6 276*	9 850 *6 121*	13 400 *8 327*	11 600 *7 208*	16 700 *10 377*	8 750 *5 437*
12 500 *7 768*	7 390 *4 592*	14 300 *8 886*	14 100 *8 762*	14 600 *9 072*	8 900 *5 530*
11 000 *6 835*	8 510 *5 288*	11 000 *6 835*	13 000 *8 078*	12 800 *7 954*	5 540 *3 443*
8 190 *5 089*	11 100 *6 898*	5 230 *3 250*	9 590 *5 959*	8 740 *5 431*	346 *215*
4 460 *2 771*	15 900 *9 880*	5 840 *3 629*	2 570 *1 597*	8 640 *5 369*	10 900 *6 773*
8 950 *5 562*	11 800 *7 333*	12 000 *7 457*	7 370 *4 580*	11 000 *6 835*	17 100 *10 626*
2 110 *1 311*	18 300 *11 372*	7 930 *4 928*	2 880 *1 790*	13 500 *8 389*	9 590 *5 959*

Shortest distances by air between 13 of the world's major cities.
Distances are shown in kilometres and *miles*.

	Los Angeles	Mexico City	New York	Paris	Singapore	Sydney
Los Angeles						
Mexico City	2 480 *1 541*					
New York	3 980 *2 473*	3 360 *2 088*				
Paris	9 090 *5 649*	9 190 *5 711*	5 830 *3 623*			
Singapore	14 100 *8 762*	16 600 *10 315*	15 300 *9 507*	10 700 *6 649*		
Sydney	12 100 *7 519*	13 000 *8 078*	16 000 *9 942*	16 900 *10 502*	6 290 *3 909*	
Tōkyō	8 800 *5 468*	11 300 *7 022*	10 870 *6 755*	9 700 *6 028*	5 290 *3 287*	7 810 *4 853*

WEB DIRECTORY

NATIONAL WEBSITES

	Official website	Tourism website
AFGHANISTAN	www.afghanistan-mfa.net	-
ALBANIA	www.keshilliministrave.al	www.albanian tourism.com
ALGERIA	www.el-mouradia.dz	www.mta.gov.dz
American Samoa	www.government.as	www.amsamoa.com/ tourism
Andorra	www.andorra.ad	www.andorra.ad
Angola	www.angola.org	www.angola.org.uk/ prov_tourism.htm
Anguilla	www.gov.ai	www.anguilla-vacation.com
ANTIGUA AND BARBUDA	www.un.int/antigua	www.antigua-barbuda.org
ARGENTINA	www.info.gov.ar	www.turismo.gov.ar
ARMENIA	www.gov.am	www.armeniainfo.am
Aruba	www.aruba.com	www.aruba.com
AUSTRALIA	www.gov.au	www.australia.com
AUSTRIA	www.oesterreich.at	www.austria-tourism.at
AZERBAIJAN	www.president.az	-
Azores	www.azores.gov.pt	www.drtacores.pt
THE BAHAMAS	www.bahamas.gov.bs	www.bahamas.com
BAHRAIN	www.bahrain.gov.bh	www.bahrain tourism.com
BANGLADESH	www.bangladesh.gov.bd	www.parjatan.org

	Official website	Tourism website
BARBADOS	www.barbados.gov.bb	www.barbados.org/bta.htm
BELARUS	www.government.by	www.mst.by
BELGIUM	www.belgium.be	Flanders: www.visitflanders.com Wallonia: www.opt.be
BELIZE	www.belize.gov.bz	www.travelbelize.org
BENIN	www.gouv.bj	www.benintourisme.com
Bermuda	www.gov.bm	www.bermudatourism.com
BHUTAN	www.bhutan.gov.bt	www.tourism.gov.bt
BOLIVIA	www.bolivia.gov.bo	-
BOSNIA-HERZEGOVINA	www.fbihvlada.gov.ba	www.bhtourism.ba
BOTSWANA	www.gov.bw	www.gov.bw/tourism
BRAZIL	www.brazil.gov.br	www.embratur.gov.br
BRUNEI	www.brunei.gov.bn	www.tourismbrunei.com
BULGARIA	www.government.bg	www.bulgariatravel.org
BURKINA	www.primature.gov.bf	www.culture.gov.bf
BURUNDI	www.burundi.gov.bi	www.burundi.gov.bi/tour.htm
CAMBODIA	www.cambodia.gov.kh	www.visit-mekong.com/cambodia/mot/
CAMEROON	www.spm.gov.cm	www.camnet.cm/mintour/tourisme
CANADA	canada.gc.ca	www.travelcanada.ca
Canary Islands	www.gobcan.es	www.gobcan.es/turismo
CAPE VERDE	www.governo.cv	-

	Official website	Tourism website
Cayman Islands	www.gov.ky	www.caymanislands.ky
CENTRAL AFRICAN REPUBLIC	-	-
Ceuta	www.ciceuta.es	-
CHAD	www.tit.td	-
CHILE	www.gobiernodechile.cl	www.visit-chile.org
CHINA	www.china.org.cn	www.cnta.com/lyen/index.asp
Christmas Island	-	www.tourism.org.cx
Cocos Islands	-	www.cocos-tourism.cc
COLOMBIA	www.gobiernoenlinea.gov.co	www.idct.gov.co
COMOROS	www.presidence-uniondescomores.com	-
CONGO	www.congo-site.com	-
CONGO, DEMOCRATIC REPUBLIC OF THE	www.un.int/drcongo	-
Cook Islands	www.cook-islands.gov.ck	www.cook-islands.com
COSTA RICA	www.casapres.go.cr	www.visitcostarica.com
CÔTE D'IVOIRE	www.pr.ci	-
CROATIA	www.vlada.hr	www.croatia.hr
CUBA	www.cubagob.gov.cu	www.cubatravel.cu
CYPRUS	www.cyprus.gov.cy	www.visitcyprus.org.cy
CZECH REPUBLIC	www.czech.cz	www.visitczech.cz
DENMARK	www.denmark.dk	www.visitdenmark.com
DJIBOUTI	-	www.office-tourisme.dj

	Official website	Tourism website
DOMINICA	www.dominica.co.uk	www.ndcdominica.dm
DOMINICAN REPUBLIC	www.presidencia.gov.do	www.dominican republic.com/Tourism
EAST TIMOR	www.gov.east-timor.org	-
ECUADOR	www.ec-gov.net	www.vivecuador.com
EGYPT	www.sis.gov.eg	www.egypt treasures.gov.eg
EL SALVADOR	www.casapres.gob.sv	www.elsalvador turismo.gob.sv
EQUATORIAL GUINEA	www.ceiba-equatorial-guinea.org	-
ERITREA	shabait.com	-
ESTONIA	www.riik.ee	visitestonia.com
ETHIOPIA	www.ethiopar.net	www.tourism ethiopia.org
Falkland Islands	www.falklands.gov.fk	www.tourism.org.fk
Faroe Islands	www.tinganes.fo	www.tourist.fo
FIJI	www.fiji.gov.fj	www.bulafiji.com
FINLAND	www.valtioneuvosto.fi	www.visitfinland.com
FRANCE	www.premier-ministre.gouv.fr	www.franceguide.com
French Guiana	www.guyane.pref.gouv.fr	www.tourisme-guyane.gf
French Polynesia	www.presidence.pf	www.tahiti-tourisme.pf
GABON	www.un.int/gabon	www.tourisme-gabon.com
THE GAMBIA	www.statehouse.gm	www.visitthegambia.qm
Gaza	www.pna.gov.ps	www.visit-palestine.com

	Official website	Tourism website
GEORGIA	www.parliament.ge	www.parliament.ge/TOURISM/
GERMANY	www.bundesregierung.de	www.germany-tourism.de
GHANA	www.ghana.gov.gh	www.ghana tourism.gov.gh
Gibraltar	www.gibraltar.gov.gi	www.gibraltar.gov.gi
GREECE	www.greece.gov.gr	www.gnto.gr
Greenland	www.nanoq.gl	www.greenland.com
GRENADA	www.grenadaconsulate.org	grenadagrenadines.com
Guadeloupe	www.cr-guadeloupe.fr	www.antilles-info-tourisme.com/guadeloupe
Guam	ns.gov.gu	www.visitguam.org
GUATEMALA	www.congreso.gob.gt	www.mayaspirit.com.gt
Guernsey	www.gov.gg	www.guernsey touristboard.com
GUINEA	www.guinee.gov.gn	www.mirinet.net.gn/ont/
GUINEA-BISSAU	-	-
GUYANA	www.gina.gov.gy	www.guyana-tourism.com
HAITI	www.haiti.org	www.haititourisme.org
HONDURAS	www.congreso.gob.hn	www.letsgo honduras.com
HUNGARY	www.magyarorszag.hu	www.hungary tourism.hu
ICELAND	eng.stjornarrad.is	www.icetourist.is

	Official website	Tourism website
INDIA	www.goidirectory.nic.in	www.tourismof india.com
INDONESIA	www.indonesia.go.id	www.budpar.go.id
IRAN	www.president.ir	www.itto.org
IRAQ	www.iraqmofa.net	-
IRELAND	www.irlgov.ie	www.ireland.travel.ie
Isle of Man	www.gov.im	www.gov.im/tourism
ISRAEL	www.index.gov.il/FirstGov	www.tourism.gov.il
ITALY	www.governo.it	www.enit.it
JAMAICA	www.jis.gov.jm	www.visitjamaica.com
JAPAN	web-japan.org	www.jnto.go.jp
Jersey	www.gov.je	www.jersey.com
JORDAN	www.nic.gov.jo	www.see-jordan.com
KAZAKHSTAN	www.president.kz	www.president.kz
KENYA	www.kenya.go.ke	www.magicalkenya.com
KIRIBATI	-	-
KUWAIT	www.kuwaitmission.com	-
KYRGYZSTAN	www.gov.kg	-
LAOS	www.un.int/lao	mekongcenter.com
LATVIA	www.saeima.lv	www.latviatourism.lv
LEBANON	www.presidency.gov.lb	www.destination lebanon.com
LESOTHO	www.lesotho.gov.ls	www.lesotho.gov.ls/ lstourism.htm
LIBERIA	www..embassyof liberia.org	-
LIBYA	www.libya-un.org	-
LIECHTENSTEIN	www.liechtenstein.li	www.tourismus.li
LITHUANIA	www.lrv.lt	www.tourism.lt

	Official website	Tourism website
LUXEMBOURG	www.gouvernement.lu	www.ont.lu
MACEDONIA (F.Y.R.O.M.)	www.vlada.mk	-
MADAGASCAR	www.madagascar-diplomatie.ch	-
Madeira	www.gov-madeira.pt/madeira	www.madeira tourism.org
MALAWI	www.malawi.gov.mw	www.tourism malawi.com
MALAYSIA	www.gov.my	tourism.gov.my
MALDIVES	www.maldivesinfo.gov.mv	www.visitmaldives.com
MALI	www.maliensdelexterieur.gov.ml	www.malitourisme.com
MALTA	www.gov.mt	www.visitmalta.com
MARSHALL ISLANDS	www.rmiembassyus.org	www.visitmarshall islands.com
Martinique	www.cr-martinique.fr	www.martinique.org
MAURITANIA	www.mauritania.mr	-
MAURITIUS	www.gov.mu	www.mauritius.net
Mayotte	-	-
Melilla	www.melilla.es	-
MEXICO	www.presidencia.gob.mx	www.visitmexico.com
MICRONESIA, FEDERATED STATES OF	www.fsmgov.org	visit-fsm.org
MOLDOVA	www.moldova.md	www.turism.md
MONACO	monaco.gouv.mc	www.monaco-congres.com

	Official website	Tourism website
MONGOLIA	www.pmis.gov.mn	www.mongolia tourism.gov.mn
Montserrat	-	www.visitmontserrat. com
MOROCCO	www.mincom.gov.ma	www.tourism-in-morocco.com
MOZAMBIQUE	www.mozambique.mz	www.mozambique. mz/turismo/topics.htm
MYANMAR (BURMA)	www.myanmar.com	www.myanmar-tourism.com
NAMIBIA	www.grnnet.gov.na	www.namibia tourism.com.na
NAURU	www.un.int/nauru	-
NEPAL	www.nepalhmg.gov.np	www.welcome nepal.com
NETHERLANDS	www.overheid.nl	www.visitholland.com
Netherlands Antilles	www.gov.an	-
New Caledonia	www.gouv.nc	-
NEW ZEALAND	www.govt.nz	www.newzealand.com
NICARAGUA	www.asamblea.gob.ni	www.visit-nicaragua.com
NIGER	www.delgi.ne/presidence	-
NIGERIA	www.nigeria.gov.ng	www.nigeriatourism.net
Niue	www.niuegov.com	www.niueisland.com
Norfolk Island	www.norfolk.gov.nf	www.norfolkisland.nf
NORTH KOREA	www.korea-dpr.com	-
Northern Mariana Islands	www.gov.mp	www.mymarianas.com

	Official website	Tourism website
NORWAY	www.norway.no	www.visitnorway.com
OMAN	www.moneoman.gov.om	www.omantourism.gov.om
PAKISTAN	www.infopak.gov.pk	www.tourism.gov.pk
PALAU	www.palauembassy.com	visit-palau.com
PANAMA	www.pa	www.visitpanama.com
PAPUA NEW GUINEA	www.pngonline.gov.pg	www.pngtourism.org.pg/
PARAGUAY	www.presidencia.gov.py	www.senatur.gov.py
PERU	www.peru.gob.pe	www.peru.org.pe
PHILIPPINES	www.gov.ph	www.tourism.gov.ph
Pitcairn Islands	www.government.pn	www.government.pn/tourist.htm
POLAND	www.poland.gov.pl	www.poland-tourism.pl
PORTUGAL	www.portugal.gov.pt	www.portugalinsite.pt
Puerto Rico	www.gobierno.pr	www.gotopuertorico.com
QATAR	english.mofa.gov.qa	www.experienceqatar.com
Réunion	-	www.la-reunion-tourisme.com
ROMANIA	www.guv.ro	www.romaniatravel.com
RUSSIAN FEDERATION	www.gov.ru	www.russiatourism.ru
RWANDA	www.gov.rw	www.rwandatourism.com
St Helena	www.sainthelena.gov.sh	www.sthelenatourism.com
ST KITTS AND NEVIS	www.stkittsnevis.net	www.stkitts-tourism.com

	Official website	Tourism website
ST LUCIA	www.stlucia.gov.lc	www.stlucia.org
St Pierre and Miquelon	-	-
ST VINCENT AND THE GRENADINES	-	www.svgtourism.com
SAMOA	www.govt.ws	www.visitsamoa.ws
SAN MARINO	www.consigliogrande egenerale.sm	www.visitsan marino.com
SÃO TOMÉ AND PRÍNCIPE	www.uns.st	www.saotome.st
SAUDI ARABIA	www.saudinf.com	www.saudi tourism.gov.sa
SENEGAL	www.gouv.sn	www.senegal-tourism.com
SERBIA AND MONTENEGRO	www.gov.yu	www.visit-montenegro.com and www.serbia-tourism.org
SEYCHELLES	www.virtualseychelles.sc	www.virtualseychelles.sc
SIERRA LEONE	www.statehouse-sl.org	-
SINGAPORE	www.gov.sg	www.visitsingapore.com
SLOVAKIA	www.government.gov.sk	www.slovakiatourism.sk
SLOVENIA	www.sigov.si	www.slovenia-tourism.si
SOLOMON ISLANDS	www.commerce.gov.sb	www.commerce. gov.sb/Tourism
SOMALIA	-	-
SOUTH AFRICA, REPUBLIC OF	www.gov.za	www.southafrica.net
SOUTH KOREA	www.korea.net	english.tour2korea.com
SPAIN	www.la-moncloa.es	www.spain.info

	Official website	Tourism website
SRI LANKA	www.priu.gov.lk	www.srilanka tourism.org
SUDAN	www.sudan.gov.sd	-
SURINAME	www.kabinet.sr.org	www.mintct.sr
SWAZILAND	www.gov.sz	www.mintour.gov.sz
SWEDEN	www.sweden.se	www.visit-sweden.com
SWITZERLAND	www.admin.ch	myswitzerland.com
SYRIA	www.moi-syria.com	www.syriatourism.org
TAIWAN	www.gov.tw	www.tbroc.gov.tw
TAJIKISTAN	www.tjus.org	www.tajiktour. tajnet.com
TANZANIA	www.tanzania.go.tz	www.tanzania touristboard.com
THAILAND	www.thaigov.go.th	www.tourism thailand.org
TOGO	www.republicoftogo.com	-
Tokelau	www.tokelau.org.nz	-
TONGA	www.pmo.gov.to	www.tongaholiday.com
TRINIDAD AND TOBAGO	www.gov.tt	www.visittnt.com
TUNISIA	www.tunisiaonline.com	www.tourism tunisia.com
TURKEY	www.mfa.gov.tr	www.turizm.gov.tr
TURKMENISTAN	www.turkmenistan embassy.org	www.turkmenistan embassy.org
Turks and Caicos Islands	-	www.turksand caicostourism.com
TUVALU	-	www.timeless tuvalu.com

	Official website	Tourism website
UGANDA	www.government.go.ug	www.visituganda.com
UKRAINE	www.kmu.gov.ua	www.tourism.gov.ua
UNITED ARAB EMIRATES	www.uae.gov.ae	-
UNITED KINGDOM	www.direct.gov.uk	www.visitbritain.com
UNITED STATES OF AMERICA	www.firstgov.gov	www.tourstates.com
URUGUAY	www.presidencia.gub.uy	www.turismo.gub.uy
UZBEKISTAN	www.gov.uz	www.uzbektourism.uz
VANUATU	www.vanuatugovern ment.gov.vu	www.vanuatu tourism.com
VATICAN CITY	www.vatican.va	-
VENEZUELA	www.gobiernoenlinea.ve	-
VIETNAM	www.na.gov.vn	www.vietnam tourism.com
Virgin Islands (U.K.)	-	www.bvitourist board.com
Virgin Islands (U.S.)	www.usvi.org	www.usvitourism.vi
Wallis and Futuna Islands	www.wallis.co.nc/ assemblee.ter	-
West Bank	www.pna.gov.ps	www.visit-palestine.com
Western Sahara	-	-
YEMEN	www.nic.gov.ye	www.yemen tourism.com
ZAMBIA	www.zambiatourism.com	www.zambia tourism.com
ZIMBABWE	www.zim.gov.zw	www.zimbabwe tourism.co.zw

GENERAL WEBSITES

COUNTRIES, POPULATION AND TRAVEL	
Countries	**Web address**
United Nations	www.un.org
European Union	europa.eu.int
Permanent Committee on Geographical Names	www.pcgn.org.uk
The World Factbook	www.odci.gov/cia/publications/factbook
Geographic Names Information System	geonames.usgs.gov
International Boundaries Research Unit	www-ibru.dur.ac.uk
World Bank	www.worldbank.org/data
Population	
UK National Statistics	www.statistics.gov.uk/census2001
City Populations	www.citypopulation.de
US Census Bureau	www.census.gov
World Urbanization Prospects	www.un.org/esa/population/publications/wup2003/WUP2003Report.pdf
United Nations Population Information Network	www.un.org/popin
UN Population Division	www.un.org/esa/population/unpc
Travel	
UK Foreign Office	www.fco.gov.uk
US Department of State	www.state.gov
World Health Organization	www.who.int
Centers for Disease Control and Prevention	www.cdc.gov/travel
Airports Council International	www.airports.org
World Wise Directory	www.brookes.ac.uk/worldwise/directory.html
Travel Daily News	www.traveldailynews.com

Theme
The United Nations
Gateway to the European Union
Place names research in the UK
Country profiles
Place names research in the USA
International boundaries resources and research
World development indicators
The UK 2001 census
Statistics and maps about population
US and world population
Population estimates and projections
World population statistics
Monitoring world population
Travel, trade and country information
Travel advice
Health advice and world health issues
Advice for travellers
The voice of the world's airports
Basic information for all countries
Travel and tourism newsletter

CLIMATE, ENVIRONMENT AND THE OCEANS AND POLAR REGIONS	
Climate	**Web address**
BBC Weather	www.bbc.co.uk/weather
The Meteorological Office	www.met-office.gov.uk
National Climatic Data Center	www.ncdc.noaa.gov
US National Hurricane Center	www.nhc.noaa.gov
National Oceanic and Atmospheric Administration	www.noaa.gov
World Meteorological Organization	www.wmo.ch
Environment	
Earth Observatory	earthobservatory.nasa.gov
National Earthquake Information Center	neic.usgs.gov
Visible Earth	visibleearth.nasa.gov
United States Geological Survey	www.usgs.gov
UNESCO World Heritage Centre	whc.unesco.org
The World Conservation Union	www.iucn.org
World Rainforest Information Portal	www.rainforestweb.org
United Nations Environment Programme	www.unep.org
World Conservation Monitoring Centre	www.unep-wcmc.org
World Resources Institute	www.wri.org
IUCN Red List	www.redlist.org
Oceans and polar regions	
International Maritime Organisation	www.imo.org
General Bathymetric Chart of the Oceans	www.ngdc.noaa.gov/mgg/gebco
National Oceanography Centre	www.soc.soton.ac.uk
Scott Polar Research Institute	www.spri.cam.ac.uk

Theme
Worldwide weather forecasts
Weather information and climatic research
Global climate data
Tracking hurricanes
Monitoring climate and the oceans
The world's climate
Observing the earth
Monitoring earthquakes
Satellite images of the earth
Volcanic activity and earthquakes
World Heritage Sites
World and ocean conservation
Rainforest information and resources
Environmental protection by the UN
Conservation and the environment
Monitoring the environment and resources
Endangered species
Shipping and the environment
Mapping the oceans
Researching the oceans
Polar research

INTERNATIONAL ORGANIZATIONS	
	Web address
United Nations	www.un.org
United Nations Educational, Scientific and Cultural Organization	www.unesco.org
United Nations Children's Fund	www.unicef.org
United Nations High Commissioner for Refugees	www.unhcr.org
European Union	europa.eu.int
Food and Agriculture Organization	www.fao.org
United Nations Development Programme	www.undp.org
North Atlantic Treaty Organization	www.nato.int
European Environment Agency	www.eea.eu.int
Europa - The European Union On-line	europe.eu.int
World Health Organisation	www.who.int
Association of Southeast Asian Nations	www.aseansec.org
Africawater	www.africawater.org
The Joint United Nations Convention on AIDS	www.unaids.org
African Union	www.africa-union.org
The Secretariat of the Pacific Community	www.spc.int
US National Park Service	www.nps.gov
The Caribbean Community Secretariat	www.caricom.org
Organization of American States	www.oas.org
The Latin American Network Information Center	lanic.utexas.edu
World Wildlife Fund	www.worldwildlife.org

Theme
The United Nations
International collaboration
Health, education, equality and protection for children
The UN refugee agency
Gateway to the European Union
Agriculture and defeating hunger
The UN global development network
North Atlantic freedom and security
Europe's environment
European Union facts and statistics
Health issues and advice
Economic, social and cultural development
Water resources in Africa
The AIDS crisis
African international relations
The Pacific community
National Parks of the USA
Caribbean Community
Inter-American cooperation
Latin America
Global environmental conservation

Main statistical sources

United Nations Statistics Division	unstats.un.org/unsd
World Population Prospects: The 2004 Revision and World Urbanization Prospects: The 2003 Revision, United Nations Population Division	www.un.org/esa/population/unpop
United Nations Population Information Network	www.un.org/popin
United Nation Development Programme	www.undp.org
World Development Indicators 2005, World Bank	www.worldbank.org/data
World Resources Institute Earth Trends Environmental Database	earthtrends.wri.org

Other Collins titles to help you expand your geographical knowledge.

Collins Complete World Atlas
A complete view of the world with maps, images and detailed content.
ISBN 0-00-720666-6

Collins need to know? The World
All the maps and facts you need to know in today's world.
ISBN 0-00-719831-0

Collins Gem World Atlas
Detailed mapping of the world in an unrivalled pocket-sized atlas.
ISBN 0-00-720561-9

Collins Gem Flags
Up-to-date guide to the flags of over 200 countries.
ISBN 0-00-716526-9

For further information visit: www.collins.co.uk